Concrete Craftsman Series
Concrete Fundamentals

CCS-0 Concrete Fundamentals was originally written by ACI Committee E703, Concrete Construction Practices.

The 2016 edition of CCS-0 was reviewed and approved by an Education Task Group including: William Nash, William Palmer, Michael Pedraza, David Suchorski, and Scott Tarr. Thank you to the Task Group members for their thoughtful and thorough review of the fundamental material.

CCS-0(16)

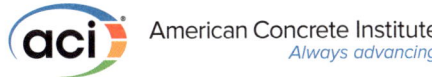

Concrete Craftsman Series:
CCS-0 Concrete Fundamentals
ISBN 978-1-942727-63-7
Copyright © 2016 American Concrete Institute

ACI Committee Reports, Guides, and Commentaries are intended for guidance in planning, designing, executing, and inspecting construction. This document is intended for the use of individuals who are competent to evaluate the significance and limitations of its content and recommendations and who will accept responsibility for the application of the material it contains. The American Concrete Institute disclaims any and all responsibility for the stated principles. The Institute shall not be liable for any loss or damage arising therefrom.

Reference to this document shall not be made in contract documents. If items found in this document are desired by the Architect/Engineer to be a part of the contract documents, they shall be restated in mandatory language for incorporation by the Architect/Engineer.

Managing Editor: Katie A. Amelio, P.E.
Art Program: Claire Hiltz
Engineering Editors: Michael Tholen, Ph.D., P.E.
   Jerzy Zemajtis, Ph. D., P.E.
Professional Development Specialist: Tiffany Vojnovski
Technical Editors: Emily Bush, Cherrie Fergusson
Manager, Publishing Services: Barry M. Bergin
Production Editors: Carl Bischof, Kelli Slayden,
   Kaitlyn Hinman, Tiesha Elam
Page Design & Composition: Ryan Jay
Manufacturing: Marie Fuller

First Printing: December 2015
Printed in Ann Arbor, Michigan.

American Concrete Institute
38800 Country Club Drive
Farmington Hills, MI 48331
USA
www.concrete.org
+1.248.848.3700

CONTENTS

PREFACE 7

VOCABULARY 9

CHAPTER 1—INTRODUCTION
What is concrete? 17
Importance of concrete 19
Importance of the craftsman 21
Adding water to concrete 21
Handling concrete safely 24

CHAPTER 2—CONCRETE MATERIALS
Portland cements 27
Types of portland cement 28
White and colored portland cement 30
Special types of cement 30
Aggregates 32
    Nominal maximum size of aggregate 34
    Aggregate grading 34
    Harmful materials in aggregate 36
    Handling aggregates 36
Mixing water 37
Admixtures 38
Supplementary cementitious materials 41

CHAPTER 3—MIXTURE PROPORTIONING
Properties of the unhardened concrete 43
Properties of hardened concrete 45
Control of shrinkage and cracking 46
Effects of temperature 47
Proportioning example 48
Values needed to choose mixture proportions 49
A summary of concrete mixture proportioning by weight
    method according to ACI 211.1 49
Concrete for the small job 55

CHAPTER 4—BATCHING AND MIXING CONCRETE

Batching	57
Mixing concrete	58
Stationary mixers: central or on site	58
Ready mixed concrete	60
Mobile batcher mixer	62
High-energy mixers	62
Remixing concrete	63
Maintenance of mixing equipment	64

CHAPTER 5—HANDLING, PLACING, AND CONSOLIDATING CONCRETE

Handling and placing methods	66
Depositing concrete from the truck mixer	66
Wheelbarrows and buggies	67
Belt conveyors	68
Buckets and hoppers	69
Pumping concrete	69
Pneumatic or air gun placing	71
Shotcrete	71
Other placing equipment and methods	72
Depositing the concrete	73
Consolidation	74
Hand methods	74
Mechanical vibration	75
Internal vibration	75
Form vibration	77
Surface vibration	77
Roller compacted concrete	78
Benefits of consolidation	78
Self-consolidating concrete	79

CHAPTER 6—CURING AND PROTECTION

Rain damage 81
Curing time and temperature 82
Accelerating admixtures 83
Keeping moisture in the concrete 84
    Membrane-forming curing compounds 84
    Waterproof paper or plastic film 85
    Water spray or soaker hose 86
    Wet burlap or mats 87
    Other methods 87
Cold weather precautions 88
    Protection against freezing 89
Hot weather precautions 91
    Keeping cool 92
    Avoiding delays 92
    Plastic shrinkage cracking 93

CHAPTER 7—FIELD TESTING AND CONTROL OF CONCRETE QUALITY

Sampling fresh concrete (ASTM C172/C172M) 95
Slump test (ASTM C143/C143M) 98
Air content tests 100
    Air content by the pressure method
    (ASTM C231/C231M) 100
    Air content by the volumetric method
    (ASTM C173/C173M) 101
    Air content estimated with an air indicator 103
Density (unit weight) and yield (ASTM C138/C138M) 104
Temperature (ASTM C1064/C1064M) 106
Making test cylinders (ASTM C31/C31M) 106
Curing and protecting test cylinders 108
    Cylinders for design strength check 109
    Cylinders made for construction site control 110

CHAPTER 8—EVALUATING CONCRETE STRENGTH

Core and Cylinder Strength Tests of Hardened Concrete	111
ACI 318 requirements	111
Cylinder compressive strength tests (ASTM C39/C39M)	112
Core tests (ASTM C42/C42M)	114
Nondestructive and in-place testing methods	115
Rebound hammer test (ASTM C805)	115
Penetration resistance method (ASTM C803/C803M)	117
Pullout tests (ASTM C900)	118
Pulse velocity test (ASTM C597)	118
Concrete maturity method (ASTM C1074)	119
Load testing concrete structures	119

APPENDIX A—REFERENCES

Referenced standards and committee reports	121
American Concrete Institute	122
ASTM International	122

PREFACE

This is one of six books in the Concrete Craftsman Series published by the American Concrete Institute. This book is intended for anyone who wants an introduction to concrete and concrete construction. Craftsmen in the concrete field may find it particularly useful as a guide for good practice.

Two other books in this series cover common concrete topics. "CCS-1 Slabs-on-Ground" covers good construction practices for slabs and is the basis for the ACI Flatwork Finisher Certification exam. "CCS-4 Shotcrete for the Craftsman" covers shotcrete construction practices and is the basis for the ACI Shotcrete Nozzleman Certification exam. CCS-4 is also available in a Spanish language edition.

Decorative concrete topics are covered in "CCS-5 Placing and Finishing Decorative Concrete Flatwork." This book provides details about the materials, equipment, and techniques required to successfully install decorative concrete flatwork.

Two other books that are part of this series are no longer in publication. "CCS-2 Cast-in-Place Walls," described formwork, reinforcement, placing of concrete, curing, and wall finishes. "CCS-3 Supported Beams and Slabs," provided technical background on such subjects as shoring, reshoring, form removal, reinforcement placement, and concrete placing, finishing, and curing.

Because this book went back to cover the fundamentals, it is numbered accordingly. "CCS-0 Concrete Fundamentals," starts with the most basic question of all, "What is concrete?" Other sections cover materials, basic construction practices, and testing. This book is a good starting point for someone in the concrete industry, whether they are an apprentice, a journeyman, a foreman, a material supplier, or even a young engineer without field experience. This book is not a design aid but rather a guide to good practice.

The design of concrete structures is the responsibility of a professional engineer. Designs are usually reviewed and approved by local building authorities and are governed by codes such as the International Building Code (IBC), or other local building codes that usually reference "Building Code Requirements for Structural Concrete (ACI 318) and Commentary." This book is not a replacement for these documents. Plans and specifications for a specific project, and local building code requirements are required to be followed, even if they differ from the information in this book.

8   CONCRETE FUNDAMENTALS

Fig. 0.1—Project plans and specifications should be followed. Practices described in this book are not a replacement for project plans

VOCABULARY

If you are using this book as an introduction to concrete, you will soon notice that a lot of ordinary words like accelerator, bleeding, blistering, honeycomb, slump, and many others have their own very special meaning when applied to concrete. To help with these meanings, listed are some of the terms that appear in this book, along with brief definitions. For longer definitions and for other words not included here, refer to "ACI Concrete Terminology (CT-13)," published by the American Concrete Institute.

Accelerator — an admixture that causes an increase in the rate of hydration of the hydraulic cement and thus shortens the time of setting, increases the rate of strength development, or both.

Admixture — a material other than water, aggregates, cementitious materials, and fiber reinforcement, used as an ingredient of a cementitious mixture to modify its freshly mixed, setting, or hardened properties and that is added to the batch before or during its mixing.

Aggregate — granular material, such as sand, gravel, crushed stone, crushed hydraulic cement concrete, or iron blast-furnace slag used with a cementing medium to produce either concrete or mortar.

Air-entraining agent — an admixture for concrete that develops a system of microscopic bubbles of air in cement paste, mortar, or concrete during mixing.

Air entrainment — the incorporation of air in the form of microscopic bubbles (typically smaller than 0.04 in. [1 mm]) during the mixing of either concrete or mortar.

Bleeding — the autogenous flow of mixing water within, or its emergence from, a newly placed cementitious mixture caused by the settlement of solid materials within the mass.

Blistering — the irregular raising of a thin layer at the surface of a placed cementitious mixture during or soon after the completion of the finishing operation (Fig. 0.2).

Bug holes — common term used to describe surface air voids. Small regular or irregular cavities, usually not exceeding 5/8 in. (15 mm) in diameter, resulting from entrapment of air bubbles in the surface of formed concrete during placement and consolidation.

Cement — any of a number of materials that are capable of binding aggregate particles together. (See also hydraulic cement.)

Cement paste — binder of concrete and mortar consisting essentially of cement, water, hydration products, and any admixtures together with very finely divided materials included in the aggregates.

Cementitious materials — pozzolans and hydraulic cements. (See also fly ash, silica fume, and slag cement.)

Clinker — a partially fused product of a kiln that is ground to make cement.

Coarse aggregate — aggregate predominantly retained on the No. 4 (4.75 mm) sieve or that portion retained on the No. 4 (4.75 mm) sieve (Fig. 0.3).

Concrete compressive strength — measured maximum resistance of a concrete specimen to axial compressive loading and expressed as a force per unit cross-sectional area.

Consolidation — the process of reducing the volume of voids, air pockets, and entrapped air in a fresh cementitious mixture, usually accomplished by inputting mechanical energy.

Craze cracks — fine random cracks or fissures in a surface of plaster, cement paste, mortar, or concrete.

Crazing — the development of craze cracks; the pattern of craze cracks existing in a surface.

Curing — action taken to maintain moisture and temperature conditions in a freshly placed cementitious mixture to allow hydraulic cement hydration and, if applicable, pozzo-

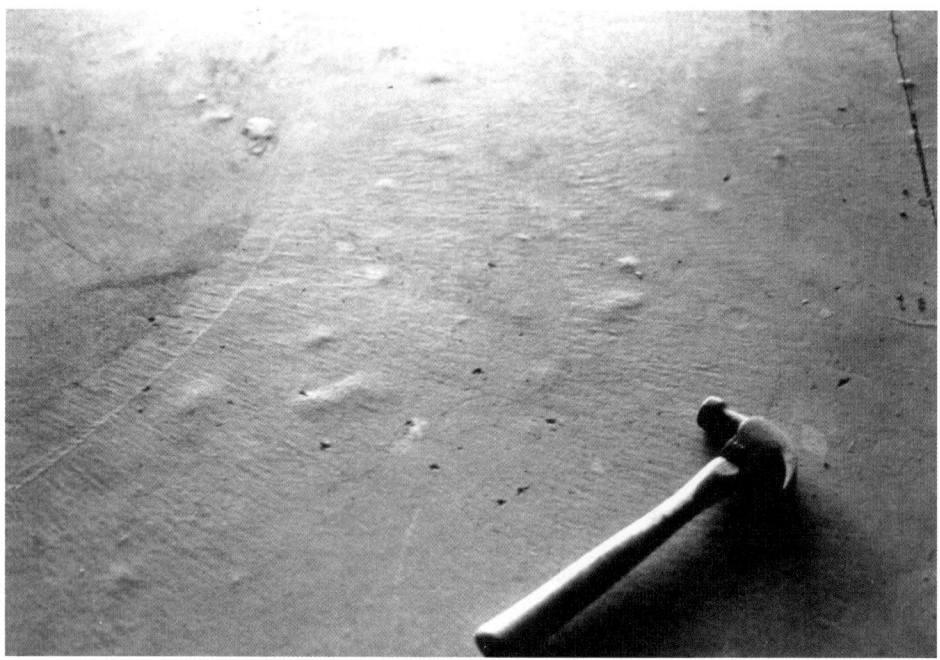

Fig. 0.2—Example of concrete surface blistering (photo courtesy of Portland Cement Association (PCA))

lanic reactions to occur so that the potential properties of the mixture may develop.

Dropchute — a device used to confine or direct the flow of a falling stream of fresh concrete. (1) *articulated dropchute* — a device consisting of a succession of tapered metal cylinders so designed that the lower end of each cylinder fits into the upper end of the one below; (2) *flexible dropchute* — a device consisting of a heavy rubberized canvas or plastic collapsible tube.

Dusting — development of a powdered material at the surface of hardened concrete.

Final setting — a degree of stiffening of a cementitious mixture greater than initial setting, generally stated as an empirical value indicating the time required for the cementitious mixture to stiffen sufficiently to resist, to an established degree, the penetration of a weighted test device. (See also initial setting.)

Fine aggregate — (1) aggregate passing the 3/8 in. (9.5 mm) sieve, almost entirely passing the No. 4 (4.75 mm) sieve, and predominantly retained on the No. 200 (75 mm) sieve; (2) that portion of aggregate passing the No. 4 (4.75 mm) sieve and predominantly retained on the No. 200 (75 mm) sieve. (See also aggregate.)

Fig. 0.3—Coarse aggregates (photo courtesy of CTLGroup)

Fineness modulus — a factor obtained by adding the total percentages of material in the sample that are coarser than each of the following sieves (cumulative percentages retained), and dividing the sum by 100: No. 100 (150 mm), No. 50 (300 mm), No. 30 (600 mm), No. 16 (1.18 mm), No. 8 (2.36 mm), No.4 (4.75 mm), 3/8 in. (9.5 mm), 3/4 in. (19.0 mm), 1-1/2 in. (37.5 mm), 3 in. (75 mm), and 6 in. (150 mm)

Fly ash — the finely divided residue that results from the combustion of ground or powdered coal and that is transported by flue gases from the combustion zone to the particle removal system.

Grout — mixture of cementitious materials and water, or other binding medium, with fine aggregate.

High-range water-reducing admixture — a water-reducing admixture capable of producing large water reduction or great flowability without causing undue set retardation or entrainment of air in mortar or concrete.

Honeycomb — voids left in concrete between coarse aggregates due to inadequate consolidation.

Hydration — the chemical reaction between hydraulic cement and water.

Hydraulic cement — a binding material that sets and hardens by chemical reaction with water and is capable of doing so underwater. For example, portland cement and slag cement are hydraulic cements.

Initial setting — a degree of stiffening of a cementitious mixture less than final set, generally stated as an empirical value indicating the time required for the cementitious mixture to stiffen sufficiently to resist, to an established degree, the penetration of a weighted test device. (See also final setting.)

Mortar — a mixture of cement paste and fine aggregate; in fresh concrete, the material occupying the interstices among particles of coarse aggregate; in masonry construction, joint mortar may contain masonry cement, or may contain hydraulic cement with lime (and possibly other admixtures) to afford greater plasticity and workability than are attainable with standard portland cement mortar.

Paste — see **Cement paste**.

Plastic shrinkage cracking — surface cracking that occurs in concrete before initial set.

Plasticity — property of freshly mixed cement paste, concrete, or mortar that determines its resistance to deformation or ease of molding.

Portland cement — a hydraulic cement made by pulverizing portland-cement clinker and usually with addition of calcium sulfate to control setting.

Pozzolans — a siliceous or silico-aluminous material that will, in finely divided form and in the presence of moisture, chemically react with calcium hydroxide at ordinary temperatures to form compounds having cementitious properties (there are both natural and artificial pozzolans).

Prestressed concrete — structural concrete in which internal stresses have been introduced to reduce potential tensile stresses in concrete resulting from loads.

Fig. 0.4—Rodding is performed using a tamping rod, the length and diameter of the rod may vary based on the size of the sample being consolidated (photo courtesy of PCA)

Retarder — an admixture that delays the setting of a cementitious mixture.

Rock pocket — a porous, mortar-deficient portion of hardened concrete consisting of coarse aggregate and voids. (See also honeycomb.)

Rodding — consolidation of concrete by means of a tamping rod (Fig. 0.4). (See also tamping.)

Scaling — local flaking or peeling away of the near-surface portion of hardened concrete or mortar (Fig. 0.5).

Segregation — the separation of coarse aggregate from the sand-cement mortar portion of the concrete mixture.

Setting — a chemical process that results in a gradual development of rigidity of a cementitious mixture, adhesive, or resin.

Fig. 0.5—Scaling removes the top layer of paste from the concrete (photo courtesy of PCA)

Sieve — a metallic plate or sheet, woven-wire cloth, or other similar device with regularly spaced apertures of uniform size mounted in a suitable frame or holder for use in separating granular material according to size.

Silica fume — very fine noncrystalline silica produced in electric arc furnaces as a byproduct of the production of elemental silicon or alloys containing silicon.

Slag cement — granulated blast-furnace slag that has been finely ground and that is hydraulic cement.

Slump — a measure of the consistency of freshly mixed concrete, equal to the subsidence of a molded specimen immediately after removal of the slump cone.

Specified compressive strength — compressive strength of concrete used in design.

Superplasticizer — See **High-range water-reducing admixture**.

Supplementary cementitious material — inorganic material such as fly ash, silica fume, metakaolin, or slag cement that reacts pozzolanically or hydraulically.

Tamping — the operation of consolidating freshly placed concrete by repeated blows or penetrations with a tamper. (See also consolidation and rodding.)

Tensile strength — maximum stress that a material is capable of resisting under axial tensile loading based on the cross-sectional area of the specimen before loading.

Tremie — a pipe or tube through which concrete is deposited under water, having at its upper end a hopper for filling and a bail for moving the assemblage.

Vibration — energetic agitation of freshly mixed concrete during placement by mechanical devices, either pneumatic or electric, that create vibratory impulses of moderately high frequency to assist in consolidating the concrete in the form or mold.

1. *External vibration* — employs vibrating devices attached at strategic positions on the forms and is particularly applicable to manufacture of precast items and for vibration of tunnel-lining forms; in manufacture of concrete products, external vibration or impact may be applied to a casting table.

2. *Internal vibration* — employs one or more vibrating elements that can be inserted into the fresh concrete at selected locations, and is more generally applicable to in-place construction.

3. *Surface vibration* — employs a portable horizontal platform on which a vibrating element is mounted.

Water-cement ratio (w/c) — the ratio of the mass of water, exclusive only of that absorbed by the aggregates, to the mass of portland cement in a cementitious mixture, stated as a decimal and abbreviated as w/c.

Water-cementitious materials ratio (w/cm) — the ratio of the mass of water, excluding that absorbed by the aggregate, to the mass of cementitious material in a mixture, stated as a decimal and abbreviated w/cm. (See also water-cement ratio.)

Workability — the property of freshly mixed concrete or mortar that determines the ease with which it can be mixed, placed, consolidated, and finished to a homogenous condition.

Yield — (1) the volume of freshly mixed concrete produced from a known quantity of ingredients; (2) the total mass of ingredients divided by the density mass of the freshly mixed concrete; (3) the number of units produced per bag of cement or per batch of concrete.

CHAPTER 1—INTRODUCTION

Those who work with concrete should know what concrete is made of and how it behaves. They should know the basic properties of concrete, and they should also recognize safety precautions needed to protect themselves and other workers when they are placing and finishing concrete.

Unlike other building materials that are delivered ready-to-use, most concrete has to be manufactured at or near the jobsite just before it is used. This makes the work of the concrete craftsman important to the success of the construction project (Fig. 1.1). An alternative is to use precast concrete that is manufactured offsite into the desired shape(s), (typically wall panels) and then transported to the jobsite (Fig. 1.2).

Understanding the basics should help workers to produce better concrete. To learn more, study the references shown in Appendix A. Three important sources, frequently referenced in this book by their initials, are:

- **American Concrete Institute (ACI):** prepares codes, specifications, guides and state-of-the-art reports for design and construction in concrete. www.concrete.org
- **ASTM International (ASTM):** prepares specifications and test methods for concrete materials and ready mixed concrete. www.astm.org
- **Portland Cement Association (PCA):** has many publications explaining how to get good quality concrete and how to build with it. www.cement.org

What is concrete?

Concrete is a mixture of cement, water, and aggregates (stone or sand) with or without admixtures, fibers, or other cementitious materials. Normalweight concrete, the concrete most commonly used for structural purposes, has a density of about 135 to 160 lb/ft^3 (2160 to 2560 kg/m^3). Lightweight structural concrete may have an equilibrium density of 90 to 115 lb/ft^3 (1440 to 1840 kg/m^3) or less, whereas special high-density (heavyweight) concrete has a

Fig. 1.1—Concrete is not delivered in its final form. The fresh concrete should be placed, consolidated, finished, and cured

Fig. 1.2—Precast panel construction

Fig. 1.3—Polished section sawed from hardened concrete. The cement-and-water paste coats each piece of aggregate and fills all spaces between the aggregate particles (photo courtesy CTLGroup)

density up to about 400 lb/ft^3 (6400 kg/m^3).

Most concrete today is made with portland cement and supplementary cementitious materials (SCMs). Concrete is sometimes described as a mixture of two major components: aggregates and paste. The paste, made of cement, SCMs, and water, binds the aggregates into a rock-like mass as the paste hardens (Fig. 1.3). The hardening is a chemical process called hydration, not a drying process, and it can take place under water as well as when exposed to air. Concrete does not harden or cure by drying, because the cement needs moisture to hydrate and harden. When the concrete dries fully it no longer gains strength.

People will often refer to the hardened mixture of cement, water, and aggregates as cement, but this is technically wrong. Concrete is the combination of these materials; only the binding powder is properly called cement (Fig. 1.4).

The paste part of the concrete also contains air, called entrapped air, which occurs when air is trapped during mixing, and is usually less than 2 percent by volume. Entrapped air voids are usually scattered, comparable in size to the larger grains of sand. Often, the paste also contains very small spherical air voids, referred to as entrained air, intentionally introduced into

CHAPTER 1—INTRODUCTION

(a)

(b)

(c)

the mixture by means of an air-entraining admixture to improve certain properties in both the fresh and hardened concrete. As Fig. 1.5 shows, the greatest part of concrete is the aggregate.

Importance of concrete

Concrete is the most widely used construction material today. Worldwide, almost 5 tons (4.5 tonnes) of concrete is produced every year for every living human being. This happens because concrete is the cheapest, most readily available material. Fortunately, it is also strong, resistant to water and fire, and readily formable to an infinite variety of sizes and shapes.

Concrete has been used to build some of the largest and tallest of all manmade structures. For example, completed in 2009, the Trump International Hotel and Tower in Chicago, IL, stands at 1389 ft (423 m). The primary structure of this building is reinforced concrete (Fig. 1.6).

Although the bulk of conventional concrete has compressive strengths in the

Fig. 1.4 — Often incorrectly used, people will refer to concrete as cement. Cement is only one ingredient in a concrete mixture. Shown in (a) is cement (powder)(photo courtesy of PCA); (b) fresh concrete (plastic state); and (c) hardened concrete (finished product)

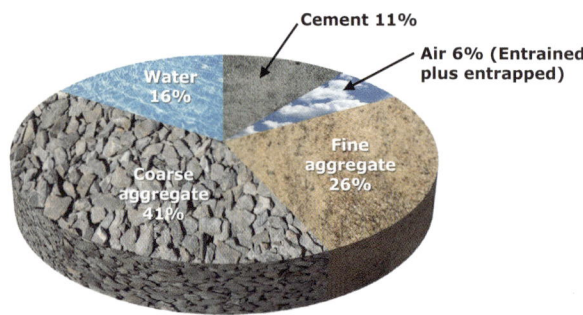

Fig. 1.5—Proportions by absolute volume of materials in a typical air-entrained concrete, freshly mixed. In other mixtures, total aggregate may range from 60 to 75 percent and cement from 7 to 15 percent

Fig. 1.6—Construction of the Trump International Hotel and Tower, Chicago, IL (photo courtesy of Lawrence Novak (PCA))

range of 3000 to 6000 psi (20 to 40 MPa), continuing advances in concrete technology are making ever-higher strengths available. Concrete for high-rise buildings frequently is designed for 9000 to 12,000 psi (60 to 80 MPa) strengths, and one building in Seattle has columns made of 19,000 psi (130 MPa) concrete. Ultra-high performance concrete (UHPC) can attain compressive strengths up to 29,000 psi (200 MPa). It is important to recognize that stronger concrete is not always better. Stronger concrete typically has a greater cement content and corresponding shrinkage potential; therefore, for thin, flat concrete structures like slabs-on-ground, stronger concrete can result in performance issues such as curling/warping and cracking. At the other end of the scale, low-strength concrete used for fill or insulation may have a compressive strength as low as 70 to 100 psi (0.5 to 0.7 MPa). One type of low strength concrete is autoclaved aerated concrete (AAC). AAC is made with fine aggregates, cement, and an expansion agent. The concrete can contain up to 80 percent air and typical compressive strengths are between 300 to 900 psi (2 to 6 MPa). AAC can be used to manufacture components that can be cut and shaped with conventional tools.

Importance of the craftsman

The concrete craftsman is a very important part of a successful concrete construction project. A successful project depends on the concrete developing the strength and other properties that the designer specified when the work was planned. The concrete quality depends directly on the concrete craftsman in the field or in a precast plant, and they should understand some of the important factors affecting how concrete develops and gains strength such as:

1. *Temperature and humidity*: This includes the temperature of the concrete as mixed and the humidity and temperature of the air it is exposed to as it sets. Warm concrete hardens faster than cool concrete but will not get as strong (lower ultimate or final strength). When humidity is low and temperature high or the wind is strong, the concrete surface can dry out causing cracks to appear before it hardens (refer to CCS-1). These cracks are commonly called plastic shrinkage cracks.

2. *Water content in relation to the amount of cementitious materials in the mixture*: Too much water causes concrete to be weaker and to shrink more.

3. *Type and fineness of cementitious materials*: The specifier usually selects one of the five standard types of portland cement described later in this book. Additionally, it is common for concrete mixtures to include SCMs such as pozzolans, slag cement, and silica fume. The choice of cement and SCMs often affects work at the jobsite because the materials can influence workability, setting time, and finishability of the concrete. Materials selection, however, is usually beyond the craftsman's control.

4. *Admixtures*: There are many types of admixtures available that are discussed in Chapter 2. Admixtures may change the setting and hardening time of the mixture. When properly selected for a particular site and weather conditions, they often make it easier for the craftsmen to do a good job. The craftsman should be aware of how admixtures may affect setting time to properly finish the concrete.

Adding water to concrete

Some concrete workers prefer the concrete mixture to be as wet as possible because this reduces the labor of placing (but not necessarily the overall labor requirement). This is not a good idea because adding water usually results in a lower strength that may not maintain compliance with specifications.

Engineers and specifiers use a term called water-cement ratio (*w/c*) to show how much water should go into the mixture. This is simply a fraction or proportion arrived at by dividing the mass of water in a batch of concrete by the mass of portland cement. A 1 yd³ (1 m³) batch of concrete containing 255 lb (150 kg) of water and 564 lb (335 kg) of cement would have

$$w/c = \frac{255 \text{ lb water}}{564 \text{ lb cement}} = 0.45 \text{ (in.-lb units)}$$

$$w/c = \frac{150 \text{ kg water}}{335 \text{ kg cement}} = 0.45 \text{ (SI units)}$$

Note: Water-cement or water-cementitious materials ratio is a dimensionless value.

Fig. 1.7—*Construction of a commercial skyscraper hotel in Dallas, TX*

Fig. 1.8—*Concrete discharging from a ready mixed concrete truck chute; each aggregate particle is coated with paste*

When concrete contains SCMs, the proper term to use is the *w/cm*, which is similar to the *w/c* except that the mass of the water is divided by the total mass of all portland cement and SCMs.

In properly made concrete, each aggregate particle is completely coated with paste (Fig. 1.8). Also, all of the spaces between aggregate particles are filled with paste. So, if the aggregates are of satisfactory quality, the quality of the concrete depends mostly on the quality of the paste. In turn, the quality of the paste depends on the proportion of water in the paste to cementitious materials. For concrete mixtures that are workable (can be placed with a reasonable amount of effort), those with low *w/cm* (small amount of water in the paste) are stronger and more durable than those with high *w/cm* (Fig. 1.9).

Most concretes have *w/cm* between about 0.4 and 0.7.

w/cm lower than 0.4 are used in low-permeability concrete and in very-high-strength concrete such as that in columns of tall buildings or dewatered specialty floor toppings. w/cm higher than 0.7 may be used in concretes where strength, durability, watertightness, and wear resistance are not critical.

While the ratio of the amount of water to the amount of cement is an important aspect in concrete mixture design, it is also important to consider the total amount of water and cement. Often, when lower w/cm are required, additional cementitious material is used as opposed to reducing the water content as intended. In these cases, even when the w/cm may be low, the mixture can have excess water.

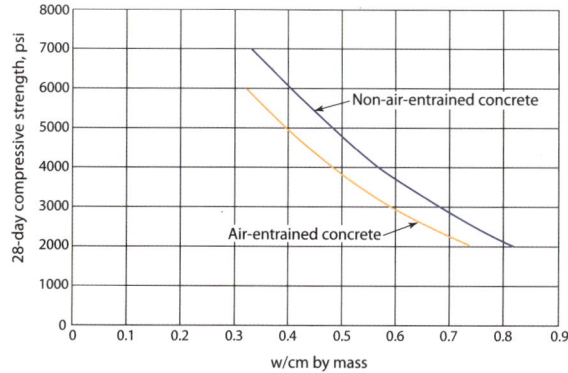

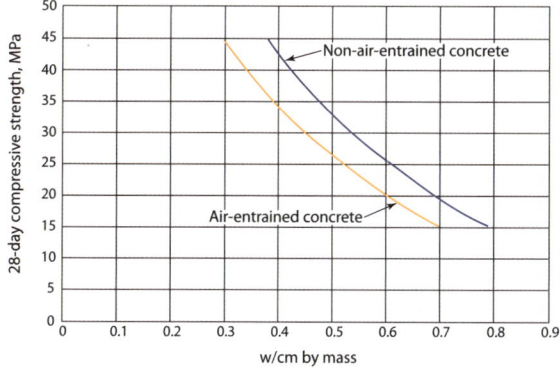

Fig. 1.9 —Strength of concrete is greater when the w/cm is low. Low w/cm values also give more durable concrete. For a given w/cm, air-entrainment causes lower strength (Values from Design and Control of Concrete Mixtures[1])

Fig. 1.10—Concrete cracking can be unsightly and depending on the location can reduce the service life of the structure

Avoiding the use of excess water will produce stronger, more durable concrete. There are other reasons to resist adding water. Concrete shrinks as it dries, and this can lead to unsightly cracking (Fig. 1.10). The more water that is used, the greater the shrinkage potential. Excess water in the concrete also increases bleeding, which is the appearance of water on the fresh concrete surface as cement grains and aggregate particles start to settle. Increased bleed water may delay finishing or cause weakness in the upper layer of the concrete. Admixtures can be used to increase workability without adding excess water. For more information, see the chemical admixture discussion in Chapter 2.

Handling concrete safely

In addition to safeguarding the concrete by avoiding too much water, workers should consider their own safety when handling concrete (Fig. 1.11). Fresh portland cement concrete is highly alkaline (caustic) and can cause skin irritation and burns. Take these simple precautions to avoid needless injury:

1. *Keep cement products off skin* — Experienced concrete craftsmen protect their skin with boots, gloves, clothing, and knee pads (Fig. 1.12). Skin injury may result from clothing that is wet from cement mixtures.

2. *Don't let skin rub against cement products* — Many cement products are abrasive. Rubbing increases the chances of serious injury. Keep concrete out of cuffs and boots.

Fig. 1.11—Hard hat, eye protection, safety vest, long-sleeved shirt, long pants, boots, and ear protection are minimum safety equipment for concrete work. Dust masks should be worn when necessary

Fig. 1.12—Worker wearing knee pads to protect his skin from fresh concrete

3. *Wash skin promptly after contact with cement products*
4. *Keep cement and cement products out of eyes* — Concrete workers should wear safety glasses or goggles. If any cement or cement mixtures get into the eye, flush immediately and repeatedly with water and consult a physician promptly.
5. *Keep products out of the reach of children* — Keep children away from cement powder and all freshly mixed cement products.

CHAPTER 2—CONCRETE MATERIALS

Freshly mixed concrete is made of cement, water, and aggregates such as sand and gravel. Frequently, it also contains one or more admixtures and supplementary cementitious materials (SCMs). The quality of these ingredients, their proportions, and the way they are mixed all affect the strength of the concrete. Typically a fresh concrete mixture should be plastic or semi-fluid, and capable of being shaped like modeling clay in the hand.

This chapter describes the following principal ingredients that make up the concrete mixture:
- Cement
- Fine aggregate
- Coarse aggregate
- Water
- Chemical admixtures
- Supplementary cementitious materials (SCMs)

Concrete as used in construction may also contain reinforcement in the form of steel or fiber-reinforced polymer (FRP) reinforcing bars (Fig. 2.1), welded wire reinforcement (sometimes called welded wire fabric or wire mesh [Fig. 2.2]), and various reinforcing fibers. Other reinforcement may be prestressed, either pretensioned or post-tensioned. Reinforcement is needed to give the concrete elements strength in tension (tensile strength). Concrete reinforcement is described in ACI document E2-00.

Fig. 2.1—Concrete workers placing concrete through reinforcement

Fig. 2.2—Welded wire reinforcement

Portland cements

Portland cements are hydraulic cements, which means they set and harden by reacting chemically with water, and they are able to do so under water. During the reaction, which is called hydration, they give off heat as they form a stone-like mass that binds the aggregate particles together. Most of the concrete's hydration and strength gain take place in the first month (typically referred to as 28 days). After that there may be slow gains in strength over long periods, sometimes 50 years or more. In massive structures such as dams and large piers, and in hot weather, the heat that is generated can cause cracking and strength loss at times. However, in most concrete work, particularly in cold weather, the heat helps concrete to harden and gain strength faster.

Origins: Portland cement was patented in 1824 by Joseph Aspdin, a mason in England. He named it "portland" because concrete made with it was similar in color to natural limestone from the Isle of Portland. Although Aspdin was the first to patent a formula for cement, natural cements produced by heating natural minerals had been used for centuries. The Greeks and Romans used lime mortars that were given hydraulic properties by the addition of volcanic ash and other natural pozzolans. The first recorded shipments of portland cement to the U.S. came from Europe in 1868, and it was first manufactured in the U.S. in 1871 (*Design and Control of Concrete Mixtures*[1]).

Manufacture: To make portland cement, finely ground raw materials such as limestone, clay, cement rock, and iron ore are blended by either a wet or dry process to produce a mixture with desired chemical composition. The raw mix is heated in a kiln to 2600 to 3000°F (1400 to 1650 °C). In the kiln it changes chemically into pellets of cement clinker. The clinker is cooled and ground to the fineness of face powder with a small amount of gypsum added to regulate the setting time. The resulting cement is not a single chemical, but is a complex mix of chemical compounds that participate in the hydration process.

Types of portland cement

ASTM C150/C150M, "Standard Specification for Portland Cement," defines the principal types of portland cement used in the U. S.:

Type I, normal cement—A general-purpose cement used where special properties of the other types are not required. Widely used in slabs-on-ground, reinforced concrete buildings, bridges, and elsewhere that concrete is not exposed to aggressive environments or to objectionable temperature rise due to heat of hydration.

Type II, moderate sulfate resistance—Used where moderate sulfate exposure is required to be resisted, as in drainage structures where sulfate in groundwater is not unusually severe. Usually generates heat of hydration more slowly than Type I. Gains strength more slowly than Types I and III, but eventually catches up (about 56 days) when concrete is properly cured.

Type III, high early strength cement—Develops strength much sooner than Type I would have at 28 days, usually in a week or less. Similar to Type I, but more finely ground so that it develops strength faster. Useful when forms need to be removed early, as for example, in precast plants or when a structure will be put into service quickly.

Type IV, low heat of hydration—Develops strength at slower rate than other types and releases less heat as it hydrates. Used in massive structures such as gravity dams and mat foundations where the rate and amount of heat generated during hardening is required to be minimized. Type IV cement is no longer manufactured in the U. S. because SCMs offer a less expensive way to control temperature rise.

Type V, high sulfate resistance—Used only in concrete exposed to severe sulfate action, principally where soils or ground waters have a high sulfate content. Sulfate attacks on concrete can cause the concrete to crack and break up unless the concrete is made sulfate-resistant.

ASTM C150/C150M also provides for three types of air-entraining cements — Types IA, IIA, and IIIA — which are similar to Types I, II, and III described previously, except that they are manufactured with small amounts of air-entraining materials that improve the resistance of concrete to freezing-and-thawing cycles. Most concrete producers believe it is more effective to add air-entraining admixtures during the concrete mixing process rather than to use special cements so there is limited demand for air-entraining cements.

Two cements that produce a moderate heat of hydration, Type II(MH) and Type II(MH)A, are also described in ASTM C150/C150M.

Type I portland cement is usually carried in stock and is supplied when the cement type is not specified. Type II is also widely available, particularly in areas with saltwater exposure or high sulfate content in the soil. Together Types I and II make up 90 percent of the cement shipped from U.S. plants. Type III and white cement are available in the larger cities, and they represent about 4 percent of cement shipments.

White and colored portland cement

The gray or tan color of ordinary portland cement depends mostly on the amount of iron in the cement. White cement, which differs from ordinary cements chiefly in color, is made to conform to ASTM C150/150M, usually Type I or Type III. White portland cement is made with selected raw materials containing very little iron or manganese oxides, which are the substances that give cement the typical gray color. White portland cement is used primarily for architectural purposes in both precast and cast-in-place concrete (Fig. 2.3). It is necessary for white concrete, and gives brighter, more intense colors for colored concrete in which pigments are added during the mixing. Colored cements are produced by intergrinding pigments with white clinker.

Special types of cement

Only 6 percent of U. S. cement production goes into special hydraulic cements. Several of the more important types of special cement are briefly described.

Blended cements: Either ASTM C595/C595M, "Standard Specification for Blended Hydraulic Cements," or ASTM C1157/C1157M, "Standard Performance Specification for Hydraulic Cement," is used to specify blended cements. Blended cements are produced by combining two or more fine materials from the following:

a.) Portland cement
b.) Slag cement
c.) Fly ash and other pozzolans
d.) Hydrated lime
e.) Ground limestone
f.) Preblended combinations of (a) through (e)

Fig. 2.3—White portland cement (photo courtesy of Lehigh White Cement Company)

The cement types produced in this way are classified by ASTM standards. The cement types in ASTM C595/C595M include:
- Type IS – Portland blast-furnace slag cement
- Type IP – Portland-pozzolan cement
- Type IL – Portland-limestone cement
- Type IT – Ternary blended cement

These cements may have optional special properties designated by the additional notation, for example, MS for moderate sulfate resistance, HS for high sulfate resistance, MH for moderate heat of hydration, LH for low heat of hydration, and more.

The cement types in ASTM C1157/C1157M include:
- Type GU – Hydraulic cement for general construction
- Type HE – High early-strength
- Type MS – Moderate sulfate resistance
- Type HS – High sulfate resistance

Blended cements can be used in construction when specific properties of these types of cements are required. However, the concrete may not gain strength as fast as with ASTM C150/C150M cements.

Masonry cements: Masonry cements are hydraulic cements manufactured to meet ASTM C91/C91M, "Standard Specification for Masonry Cement," requirements designed for use in mortar for masonry construction. They contain a variation of cement or hydraulic lime, usually in combination with other materials such as hydrated lime, limestone, chalk, slag, or clay. There are three types— M, N, and S—that are used with or without other cements to obtain workable, plastic masonry mortars. Masonry cements can be used for parging and plaster (stucco), but never for making concrete.

Expansive cements: These hydraulic cements expand rather than shrink during early hydration after setting. Manufactured to meet ASTM C845/C845M, "Standard Specification for Expansive Hydraulic Cement," they are used most commonly to produce shrinkage-compensating concrete. There are three classifications—K, M and S. When the expansion is properly restrained, they are effective in making crack-free pavements and slabs, as well as plugging leaks in concrete and masonry walls. Expansive cements are not commonly manufactured; most concrete producers believe it is more economical to add an expansive admixture during the concrete mixing process rather than use a special expansive cement.

In addition, there are also oil-well, refractory, plastic, waterproof, and regulated-set cements, as well as others still under development. Refer to *Design and Control of Concrete Mixtures*[1] for more information about blended cements.

Aggregates

The sand, gravel, crushed stone, and similar materials that are mixed with cement and water to make concrete are called

Fig. 2.4—Principal use of white portland cement is in architectural concrete, both precast and cast-in-place. The Advocate Condell Medical Center was renovated with architectural precast using white portland cement (photos courtesy of Condell Medical Center)

aggregates. These materials make up 60 to 75 percent of the absolute (solid) volume of concrete and represent 70 to 80 percent of its mass. A 1 yd^3 (1 m^3) batch of normalweight concrete can contain 2600 to 3200 lb (1500 to 1900 kg) of aggregates.

Fine aggregate: If all of the particles are smaller than 3/8 in. (9.5 mm), the aggregate is defined as fine. Although most fine aggregates are natural sand, some are produced by crushing rock.

Coarse aggregate: If most of the particles are larger than about 1/4 in. (6.4 mm) the aggregate is defined as coarse. It can be either gravel or crushed material. Gravel usually has smoothly rounded particles, while crushed stone has rough, angular surfaces. The largest pieces of coarse aggregate can measure 6 in. (150 mm) or more.

Concrete that is made without coarse aggregate is usually called mortar or grout. Most concrete used in building construction has a maximum aggregate size from 3/4 to 1-1/2 in. (19 to 37.5 mm). In massive structures like dams and mat foundations, larger aggregate sizes are often used.

Natural gravel and sand are usually taken from a pit, river, lake, or seabed. Aggregates from saltwater areas should be used with caution because of problems with seashell and salt content. Crushed aggregate can be made from natural gravel, quarry rock, cobbles, or boulders. Air-cooled blast-furnace slag is also crushed to make both fine and coarse aggregate. Recycled or waste concrete can also be crushed to make satisfactory aggregate. Whether the aggregate is crushed or natural, the mix of particle sizes is not always ideal for making concrete. The aggregate producer may pass aggregate over screens that sort out particles of various sizes, then recombine them in proportions given in standard specifications.

The most common aggregates such as sand, gravel, crushed stone, or crushed slag, make concretes with densities from 135 to 160 lb/ft^3 (2160 to 2560 kg/m^3) when freshly mixed. Structural lightweight concrete with density from about 90 to 120 lb/ft^3 (1440 to 1920 kg/m^3) is made with aggregates of expanded shale, fired clay, slate, or slag. Even

Fig. 2.5—Technician grading coarse aggregate (photo courtesy of PCA)

lighter aggregates such as pumice, perlite, and vermiculite can be used to produce concretes with densities from 15 to 90 lb/ft^3 (240 to 1440 kg/m^3). At the other extreme, heavy materials such as barite, limonite, and iron and steel punchings are used in radiation-shielding concretes weighing up to 380 lb/ft^3 (6100 kg/m^3). Only the normalweight concrete and aggregates are reviewed here. Refer to *Design and Control of Concrete Mixtures*[1] for more information about other concretes.

Normalweight aggregates should meet the requirements of ASTM C33/C33M, "Standard Specification for Concrete Aggregates." A number of tests can be made to determine if an aggregate can be used to make strong durable concrete. Important factors that affect the quality of the concrete are:
1. Nominal maximum size
2. Grading
3. Particle shape
4. Organic impurities
5. Silt and clay content
6. Amount of coarse and fine aggregate in the mix

Nominal maximum size of aggregate

The largest aggregate that can be used depends on the size and shape of the member being made from the concrete, and on the spacing and location of reinforcing steel in it. Size of the largest pieces of aggregate are not allowed to be more than:
1. One-fifth of the narrowest dimension of the concrete member
2. Three-fourths of the smallest clear distance between reinforcing bars or bar bundles and the space between bars and form
3. One-third the depth, if concrete is going into a slab

Aggregate grading

Aggregates are made up of particles of many different sizes. To keep on making concrete batches that are essentially the same, the aggregate amount and distribution of particle sizes need to be essentially the same. To measure the particle sizes, a dry sample of the aggregate is passed through a number of standardized sieves (Fig. 2.6) starting with the largest openings and using smaller and smaller openings in successive sieves. The grading can then be defined precisely by the percentage of the total mass passing each sieve. Called a sieve analysis, refer to ASTM C136/C136M, "Standard Test Method for Sieve Analysis of Fine and Coarse Aggregates," for this process.

Fig. 2.6—Sieves are available in different size diameters; larger sieves (Fig. 2.5) are available for larger quantities of aggregate. Many sieves are available for use in a mechanical shaker

There are several reasons why ASTM C33/C33M specifies grading limits and maximum aggregate sizes. The amount of cement and water needed for a mixture depends on the aggregate size. For example, with 1 in. (25 mm) maximum size aggregate, less cement and water are needed to produce concrete with a given strength than with 1/2 in. (12.5 mm) aggregate. Workability, porosity, resistance to freezing-and-thawing, pumpability, and other properties can also be affected by the grading and aggregate size. Variations in grading can also seriously affect the uniformity of concrete from one batch to the next. In general, aggregates that do not have a large deficiency or excess of aggregate passing each of the sieves will produce the best results. These aggregates are said to have a smooth grading curve. The specific grading selected is less critical than the need to prevent variation from batch to batch. The grading limits in ASTM C33/C33M are very wide to accommodate the variation found throughout the country. However, the variation within a specific source should be much smaller. The higher the tolerance in the finish, such as with superflat floors, requires more specific gradation and less variability. The gradation of the aggregates can also be optimized to result in a minimized amount of cement paste and corresponding minimal shrinkage for that specific aggregate source. However, the aggregate source has one of the greatest influences on concrete shrinkage

Harmful materials in aggregate

Most specifications limit the amount of potentially harmful substances in aggregates (Table 2.1). ASTM C33/C33M places maximum allowable percent limits on deleterious materials, such as clay lumps and friable particles, in fine and coarse aggregates. A primary concern is that poor aggregates will harm the concrete durability. Additionally, organic impurities such as peat and humus may delay setting and hardening of concrete and reduce its strength gain. Many impurities can cause popouts, which are the breaking away of small parts of a concrete surface due to localized internal pressure. Typically, they are caused by freezing of moisture in unsound aggregates, and leave cone-shaped fractures in the concrete surface (Fig. 2.7). In other examples, impurities are expansive without freezing, such as lignite; this can result in numerous flakes in the surface.

Handling aggregates

All handling and stockpiling operations cause some segregation and breakage of coarse aggregates. Crushed aggregates segregate less than rounded ones, and larger aggregates segregate more than smaller ones. Sometimes the different size fractions of

Table 2.1—Harmful Substances in Aggregates

Substances	Effect on concrete
Organic impurities	Affects setting and hardening, may cause deterioration
Materials finer than the No. 200 (75 mm) sieve	Affects bond, increases water requirement
Coal, lignite, or other lightweight materials	Affects durability, may cause stains and popouts
Soft particles	Affects durability
Clay lumps and friable particles	Affects workability and durability, may cause popouts
Chert of less than 2.40 relative density	Affects durability, may cause popouts
Alkali-reactive aggregates	Causes abnormal expansion, map cracking, and popouts

Note: Information is from *Design and Control of Concrete Mixtures*[1] Table 5-6.

Fig. 2.7—A popout occurs when internal pressure causes a small part of the concrete surface to break away, usually leaving a cone-shaped hole. Popouts range in size from 1/4 in. (6 mm) or less in diameter to 2 in. (50 mm) or more, depending on aggregate size (photo courtesy of Frances Griffith)

Fig. 2.8—Aggregate stockpiles should be kept clean and separate. Handling to minimize segregation is important (photo courtesy of PCA)

coarse aggregates are stockpiled and batched separately. Proper stockpiling procedures, however, can at times eliminate the need for this (Fig. 2.8).

When aggregates are delivered by truck, the most economical and acceptable method is the truck-dump method, which discharges the loads in a way that keeps them tightly joined; particles that roll down the sides of the pile cause segregation. The aggregate is later moved with a front-end loader that should slice off the edges of the pile from top to bottom so that every slice contains a portion of each horizontal layer. *Design and Control of Concrete Mixtures*[1] gives detailed information on handling and storage of aggregates. Where ready mixed concrete is used, the producer routinely handles aggregate storage methods.

Mixing water

Water quality is a concern because chemicals in it, even in very small amounts, sometimes change the setting time of the mixture or the long-term properties of the concrete. Almost any natural water that is drinkable can be used to make concrete. Treated city water, or tap water, from most U.S. and Canadian cities with a populations of 20,000 or more are also suitable for making concrete. Some water that is not drinkable, including recycled waters from mining and other industrial operations, may also be used, but tests should be made of such water before using it. Non-drinkable water should be tested in accordance with ASTM C1602/C1602M, "Standard Specification for the Mixing Water Used in the Production of Hydraulic Cement Concrete."

Seawater containing up to 35,000 parts per million (ppm) of dissolved solids, is not harmful to the strength of plain (unreinforced) concrete, but it may cause reinforcement to rust in reinforced concrete. Waters containing appreciable amounts of oil, salt, algae, or sugar should be checked before using in concrete. Mortar cubes made with questionable water should have at least 90 percent of the reference strength of specimens made with clean water. Follow ASTM C191, "Standard Test Methods for Time of Setting of Hydraulic Cement by Vicat Needle," to assure that the water does not affect the setting time of the cement to an unacceptable degree. Where specifications place severe limitations on chloride content of concrete, it might be necessary to consider the chloride content of mixing water when selecting concrete proportions.

Admixtures

Any material deliberately added to change the fresh or hardened characteristics of concrete, before or during mixing (Fig. 2.9) other than cementitious materials, water, aggregates and fiber reinforcement is called an admixture. There are many kinds of admixtures, including:

- Air-entraining admixtures
- Accelerating admixtures
- Retarding admixtures
- Water-reducing admixtures
- High-range water-reducing admixtures (superplasticizers)
- Miscellaneous special purpose admixtures, such as colors, corrosion inhibitors, and pumping aids

ASTM C260/C260M, "Standard Specification for Air-Entraining Admixtures for Concrete," is a standard specification specifically for air-entraining admixtures. ASTM C494/C494M, "Standard Specification for Chemical Admixtures for Concrete," defines requirements for accelerating admixtures, retarding admixtures, water-reducing admixtures, and high-range water-reducing admixtures (superplasticizers). It also provides for dual-function admixtures that combine water reduction with either acceleration or retardation of the mixture. Finely divided SCMs such as fly ash and natural pozzolans are covered by ASTM C141/C141M, "Standard

Fig. 2.9—Admixture dispensing into a concrete mixture

Specification for Hydrated Hydraulic Lime for Structural Purposes," ASTM C618, "Standard Specification for Coal Fly Ash and Raw or Calcined Natural Pozzolan for Use in Concrete," and ASTM C989/C989M, "Standard Specification for Slag Cement for Use in Concrete and Mortars."

Air-entraining admixtures produce tiny air bubbles in concrete (Fig. 2.10). The bubbles are formed by the mixing action and the air-entraining admixture keeps the bubbles from breaking up. Do not confuse entrained air with ordinary, larger air bubbles trapped in the concrete during mixing. The largest of the purposely entrained bubbles are 4/100 of an inch (1 mm) in diameter; the smallest ones range down to 4/10000 of an inch (10 μm), no larger than the width of a delicate hairline crack. The bubbles should be small because in addition to the amount of entrained air, the bubbles need to be closely spaced to provide the intended benefit. These purposely entrained air bubbles are uniformly distributed throughout the concrete, giving it greatly improved ability to withstand freezing and thawing. The air bubbles in the paste provide room for expansion of the freezing water, thus relieving the otherwise disruptive pressure. Air-entrainment also makes fresh concrete more workable for a given water content and helps reduce bleeding and segregation.

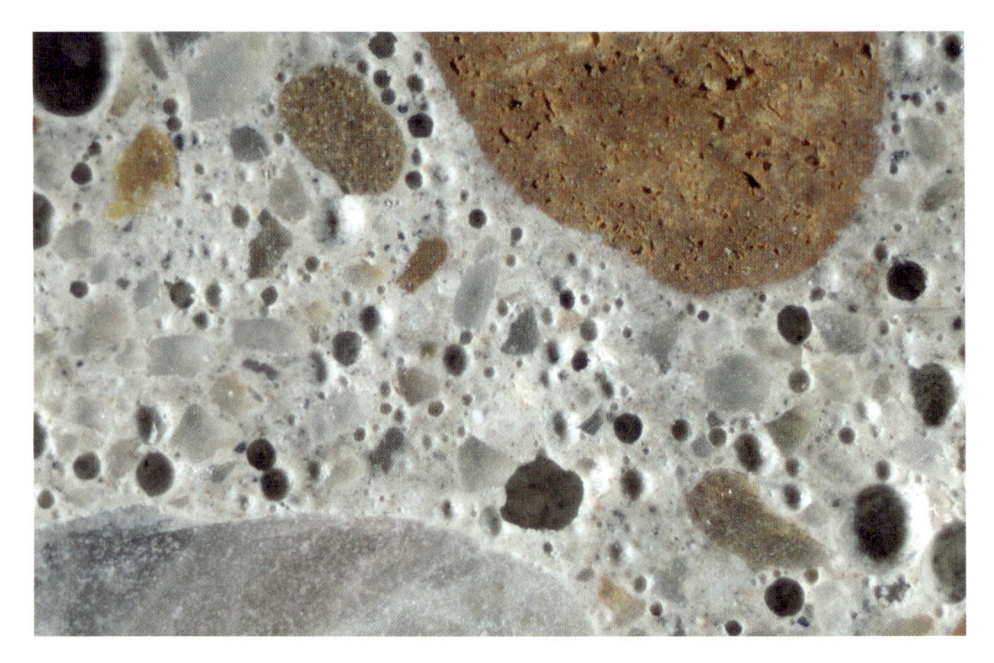

Fig. 2.10—View of air-void system in hardened concrete. The tiny air bubbles are maintained in the hardened structure by the use of air-entraining admixtures (photo courtesy of CTLGroup)

Accelerating admixtures speed up the setting and hardening of concrete. They are especially useful in cold weather because concrete hardens slowly at temperatures below about 50°F (10°C). The most common of these admixtures is calcium chloride. However, calcium chloride in concrete can increase the potential for corrosion of reinforcing steel and some other metals. When required by the specifications, non-chloride accelerators are available for use. Accelerators do not, however, replace proper curing and frost protection. The intent of accelerating admixtures is to promote the same setting time and bleed period for concrete placed during warmer temperature. When used during warm temperature, accelerators shorten the setting time and bleed period, which can result in an increase in the shrinkage potential and corresponding risk of cracking.

Retarding admixtures slow down the initial setting of the concrete. They are often used in warm weather to keep the concrete from setting before it can be placed and finished. Most retarding admixtures are also water-reducing admixtures. And, just as accelerators should not be used in warm weather, using a retarder during colder temperatures could delay the setting time and prolong the bleed period too much. As with accelerators in cold weather, the purpose of a retarder is to promote a set time and bleed period during warm weather similar to that during normal placement temperatures.

Water-reducing admixtures, as the name suggests, reduce the amount of water needed to produce a cubic yard of concrete of a given slump. If they are added to a mixture without reducing the amount of water, water reducers will act to increase the slump of the concrete.

High-range water-reducing admixtures (ASTM C494/C494M and ASTM C1017/C1017M, "Standard Specification for Chemical Admixtures for Use in Producing Flowing Concrete") are commonly referred to as "superplasticizers." They reduce the water requirement for concrete dramatically. They can be used to temporarily increase the slump or flowability of normal or stiff concrete to make it more flowable and easier to place with little vibration. They can also be used to reduce the water needed while improving workability of concretes that are consolidated by vibration. The action of some superplasticizers lasts only about 30 to 60 minutes at normal temperatures. Then the concrete stiffens very rapidly. For this reason, they are often added to concrete at the jobsite. However, caution is advised when adding a chemical at the jobsite to ensure the setting behavior does not vary between batches. Extended life superplasticizers are available for addition at the batch plant.

Supplementary cementitious materials

Supplementary cementitious materials (SCMs) may be used in large amounts, generally in the range of 20 to 100 percent replacement by mass of portland cement. They are powdered or pulverized siliceous materials added before or during mixing to improve or change the properties of concrete. They are generally natural or by-product materials such as:
- Hydraulic hydrated lime (ASTM C141/C141M)
- Slag cement (ASTM C989/C989M)
- Fly ash and natural pozzolans (ASTM C618)
- Silica fume (ASTM C1240)

Some SCMs, such as lime and slag cement, have cement-like properties—that is, they can set and harden in the presence of water. Pozzolans by definition are materials that have little cementitious action when used alone, but when used with portland cement they react with products of cement hydration to develop additional cementing action. These reaction products are the same as those needed for silicate-based surface hardeners, which can be less effective when applied to concrete containing fly ash. However, some of the finely divided minerals have both cementitious and pozzolanic properties.

Today the most commonly used pozzolan is fly ash, a finely divided residue that results from burning ground or powdered coal, which is usually the by-product of coal-burning power plants. Fly ash particles are spheres with typical diameter less than 20 millionths of an inch (500 millionths of a millimeter), somewhat finer than many portland cement particles. Some fly ash has both pozzolanic and cementitious properties.

Condensed silica fume, sometimes called microsilica, has gained attention recently as a pozzolan with particles only one hundredth the size of fly ash particles. Silica fume is a byproduct of induction arc furnaces in the silicon metal and ferrosilicon alloy industries.

Pozzolans as admixtures or replacement for part of the portland cement in a mixture improve the workability of fresh concrete; reduce thermal cracking in massive structures, because they reduce the heat of hydration; and by reducing the permeability of concrete, improve its durability. In the amounts normally used, most pozzolans cause a reduction of early strengths up to 28 days, but improve the long-term strength. Some authorities (*Design and Control of Concrete Mixtures*[1]) consider use of pozzolans essential to production of high strength concretes above 8500 psi (60 MPa).

CHAPTER 3—MIXTURE PROPORTIONING

Choosing the proper amounts of ingredients to make a batch of concrete is referred to as mixture (or mix) proportioning, or more commonly mixture design. The ideal amount of cementitious materials, water, and aggregates required to produce a cubic yard (cubic meter) of concrete can be determined, usually from a series of trial batches, in such a way to meet the following objectives:
- The freshly mixed concrete is placeable, workable, and finishable for the specific environmental conditions.
- The hardened concrete has the strength, durability, and uniform appearance called for within the project specifications.
- The mixture is economical.

Although the craftsman may not have responsibility for mixture proportioning, it is a good idea to understand some of the principles and practices involved. The actual mixture proportioning is usually done by the concrete producers or by independent testing laboratories and the engineers and technicians working for them. They frequently can draw on long experience or familiarity with their locally sourced concrete-making materials. In many cases, the job specifications set requirements of such properties as strength, air content, and slump. Sometimes other limits or properties are specified; for example, cement content, admixture type and dosages, aggregate size, and pumpability.

Desirable properties of the freshly mixed concrete as well as the hardened concrete are outlined here to show the many factors that have a bearing on the mixture design.

Properties of the unhardened concrete

Workability of concrete is the ease with which concrete can be placed, consolidated, and finished without causing harmful segregation. It is difficult to measure workability, but craftsmen who work with concrete can judge if one mixture is more workable than another. *Consistency* is the ability of freshly mixed concrete to flow. *Plasticity* indicates the concrete's ease of molding. The slump test (Fig. 3.1), performed on concrete as delivered, is used to measure and control consistency or flowability.

Bleeding occurs as the concrete in place begins to harden. Some of the aggregate particles and the cement grains are partly suspended in water so they tend to settle. As these solids settle, water is displaced and appears at the surface. This water is called *bleed water* and the process is called *bleeding*.

Fig. 3.1—Slump measured after removal of the cone-shaped container indicates concrete consistency. The test, described in Chapter 7, "Slump test (ASTM C143/C143M)" section, is used routinely to check batch-to-batch uniformity of ready mixed concrete (Manual of Concrete Inspection[2]) (photo courtesy of Frances Griffith)

The amount of bleed water that appears on the surface of fresh concrete depends on how the mixture proportions were selected. More bleeding can be expected with very wet concrete mixtures, with non-air-entrained mixtures, and in mixtures that contain little fine sand. Lean mixtures with low cement content will bleed more than rich mixtures. Thus, the mixture design determines the amount of bleeding, which significantly affects the work of the concrete finisher who should wait for bleed water to disappear before finishing a slab. Finishing before the bleeding stops or while bleed water remains on the surface may cause the slab to dust or scale during its service life. Allowing the bleed water to leave the slab before the final set reduces the *w/cm* ratio of the concrete and improves its properties. Reducing the amount of bleed water by using an accelerating admixture results in more free water retained in the concrete, thus, potentially increasing the drying shrinkage behavior and corresponding cracking, warping, and drying time. When an accelerating admixture is used, it is important to maintain consistent setting time and bleed period behavior.

Properties of hardened concrete

Compressive strength (Fig. 3.2) is said to be inversely related to the *w/cm*, within the normal range of concrete strengths. That is, when the *w/cm* is lower, the strength can be expected to be higher, for a given set of materials. The w/cm, as explained in Chapter 1, "Adding water to concrete" section, is simply the mass of water divided by the mass of cement in a given batch of concrete. What the ratio should be to produce a given strength of concrete can be estimated from tests of concrete made with different *w/cm*. Increasing the amount of water without increasing the amount of cement will result in lower strengths when other parts of the mixture are kept the same.

The compressive strength values used in determining mixture proportions for the start-up of concreting are not the same as f'_c, the specified compressive strength (usually at 28 days age) given by the architect/engineer in the specifications and used in the design calculations. Codes require a certain degree of strength overdesign, using statistical methods to minimize the probability of getting strengths lower than f'_c on the job. When the concrete producer has a suitable record of tests of concrete using similar materials under similar conditions, the required overdesign may be only a few hundred psi (MPa). Where there are insufficient test results for the materials and mixture conditions, the required overdesign ranges from 1000 to 1400 psi (7 to 10 MPa).

Durability is the ability of a material to resist weathering action, chemical attack, abrasion, and other conditions of

Fig. 3.2—Technician is centering a 6 x12 in. (150 x 300 mm) cylinder in a load machine for compressive strength testing (photo courtesy of PCA)

service including resistance to freezing and thawing. Air-entrained concrete is able to withstand cycles of freezing and thawing much better than non-air-entrained concrete. Air-entrainment should be used for all concrete that is exposed to freezing. Durability may also mean resistance to abrasion or chemical attack. Here, high strength and low *w/cm* are also important.

Watertightness of a particular concrete mixture depends primarily on the strength, the *w/cm* used, air content, and duration of curing. It requires a nonporous aggregate, with each particle surrounded by dense paste. The mixture should be proportioned to provide workable concrete that will not segregate. Avoid harsh mixtures which tend to form rock pockets (clumps of stone with open voids). Use of pozzolans and silica fume in the mixture also helps in attaining watertightness.

Curing of concrete is action taken to maintain satisfactory temperature and moisture conditions during the first few days after placing. Curing is essential to development of the strength, durability, and watertightness of the finished construction. Ideally, curing should start immediately after concrete placement and finishing.

Control of shrinkage and cracking

As new concrete dries, it shrinks, or shortens, in all directions. Most concretes shrink as much as 1/8 in. per 20 ft (0.5 mm/m). Exposed surfaces dry faster than the rest of the concrete. Additional shortening occurs when concrete cools.

Concrete is strong in compression, but weak in tension. When concrete shrinkage is resisted by friction between the concrete element and its supports, tension forces develop, and when they exceed the tensile strength of the concrete, it cracks. Designers provide joints or reinforcing steel to control the cracking. This is explained in "CCS-1 Slabs-on-Ground." To control the shrinkage itself, the most important factor in shrinkage is the total amount of water per cubic yard.

To reduce shrinkage and thereby reduce cracking, apply the following rules:

1. Specify a maximum *w/cm*, and use the stiffest mixture that can be placed and consolidated properly.

2. Use the largest nominal maximum size aggregate that is practical. Concrete made with 3/8 in. (9.5 mm) maximum aggregate usually requires about 40 lb (nearly 5 gallons) more water per cubic yard (14 kg (about 18 L) more water per cubic meter) to get the same slump that concrete made with 1 in. (25 mm) maximum size aggregate does.

Fig. 3.3—Concrete skate park in Malmo, Sweden. Concrete can be used to build unique landscape, park, and building structures (photo courtesy of Jerzy Zemajtis)

3. Use water-reducing, set-controlling admixtures (ASTM C494/C494M) that show good performance in shrinkage tests.

Use of clean, coarse sand and Type II cement also can help to reduce shrinkage.

Effects of temperature

The hardening of concrete speeds up if its temperature is warm and slows down when cool. For example, if a concrete mixture begins to set in 2 hours at 70°F (21°C), it may set in an hour or less at 95°F (35°C). For the same mixture, the setting time may increase to 3 hours or more at 50°F (10°C) and 5 hours or longer at 35°F (2°C).

Temperature also affects the amount of water needed to make a cubic yard (meter) of workable concrete. At low temperatures, less water is needed to get a workable mixture than at high temperatures. For typical concrete mixtures having a given slump, an additional gallon of water per cubic yard (5 L of water per cubic meter) of concrete is needed for each 12°F (7°C) increase in its temperature. This means that if the same amount of cement is used (per yd^3 (m^3) of concrete), the *w/cm* and total water per cubic yard (meter) is higher for warm concrete than for cold concrete with the same slump. The warm concrete will also shrink more.

What is the best temperature for making concrete? The answer depends on how long the concrete is expected to be cured. If concrete is to be cured naturally for long periods, such as in dams or foundations, concrete made at cold temperatures just above freezing will eventually be stronger than concrete made at higher temperatures. Concrete that can be cured only a few days, however, should be kept above 60°F (16°C). Heat of hydration generated by the setting action of cement also affects concrete strength gain.

Concrete that freezes soon after it is made and before its strength reaches about 500 psi (3.5 MPa) may be permanently damaged. Frozen concrete should be removed and replaced. Admixtures do not lower the freezing point of concrete significantly; therefore, admixtures do not behave as anti-freeze agents. Accelerating admixtures, however, will speed up hardening and reduce the time required to obtain a strength of 500 psi (3.5 MPa). Air-entrainment will improve performance of concrete subject to freezing and thawing during construction.

Proportioning example

To produce concrete that is economical but still has the required strength, durability, and workability and other properties just described, use the least amount of cementitious materials that will give the required *w/cm* and the most coarse aggregate that can be used and still have a workable, placeable, and finishable mixture.

New mixtures are commonly proportioned by individuals or laboratories that specialize in concrete testing and control because laboratory tests are needed to verify the proportions. Two proportioning procedures commonly used in the U.S. are the *weight method* and the *absolute volume method*. Both are explained in ACI 211.1, "Standard Practice for Selecting Proportions for Normal, Heavyweight, and Mass Concrete."

CHAPTER 3—MIXTURE PROPORTIONING

A summary of concrete mixture proportioning by weight method according to ACI 211.1

Job specifications and codes establish many requirements for concrete. Read and understand them. The average strength for which the concrete is proportioned will be more than the specified strength. For guidance refer to Section 4.2.3 of ACI 301, "Specifications for Structural Concrete."

1. Select the slump...either as specified or use Table 6.3.1 of ACI 211.1.
2. Choose the maximum aggregate size using the largest economically available aggregate that meets the specifications and code limits of ACI 318, Section 26.4.2.1a.
3. Estimate mixing water and air content using Table 6.3.3 of ACI 211.1. (Refer to Table 3.1 in this book for a simplified version.) Use specified air content if given; otherwise estimate.
4. Choose admixtures to meet specifications and job conditions.
5. Choose w/cm, either as specified or based on the strength required, using Table 6.3.4a of ACI 211.1. (Refer to Table 3.2 in this book for a simplified version.)
6. Calculate cement content based on estimated water and the w/cm found in Steps 3 and 5.
7. Estimate volume of coarse aggregate using Table 6.3.6 of ACI 211.1. (Refer to Table 3.3 in this book for a simplified version.)
8. Find mass of coarse aggregate using data on dry-rodded density of the aggregate being used.
9. Estimate mass of fine aggregate by subtracting total mass of other ingredients from estimated mass-per-unit volume of concrete. Use known mass from previous experience or follow Table 6.3.7.1 of ACI 211.1. (Refer to Table 3.4 in this book is a simplified version.)

This procedure gives mass of materials for a cubic yard (meter) of concrete. Scale down to a trial batch, make specimens and test for strength and other specified properties. Adjust proportions as needed, make another batch, and retest. Several trials may be needed.

ACI 211.1 gives tables and data for general guidance in proportioning, and these are used in the example which follows. However, it is preferable to use data derived for the actual ingredients and field conditions whenever they are available. The steps outlined by ACI 211.1 are shown in the following example using the weight method of proportioning.

Values needed to choose mixture proportions

Assume that a mixture is needed for a 5 in. (125 mm) unreinforced concrete slab-on-ground. Some requirements have been set, as follows:

Maximum w/cm = 0.52

Table 3.1—Approximate Mixing Water and Air Content Requirements for Air-Entrained Concrete of Different Slumps and Nominal Maximum Sizes of Aggregates (ACI 211.1)

SLUMP	WATER, lb/yd³ (kg/m³) for concrete made with indicated nominal maximum size of aggregate				
	1/2 in. (12.5 mm)	3/4 in. (19 mm)	1 in. (25 mm)	1-1/2 in. (37.5 mm)	2 in. (50 mm)
1 to 2 in. (25 to 50 mm)	295 (175)	280 (166)	270 (160)	250 (148)	240 (142)
3 to 4 in. (75 to 100 mm)	325 (193)	305 (181)	295 (175)	275 (163)	265 (157)
6 to 7 in. (150 to 175 mm)	345 (205)	325 (193)	310 (184)	290 (172)	280 (166)
EXPOSURE	Recommended average total air content (percent) for exposure level shown				
MILD	4.0	3.5	3.0	2.5	2.0
MODERATE (F1 in ACI 318)	5.5	5.0	4.5	4.5	4.0
SEVERE (F2 and F3 in ACI 318)	7.0	6.0	6.0	5.5	5.0

Note: MODERATE exposure level in Table 3.1 corresponds to Exposure Category F1 in ACI 318 and SEVERE exposure level corresponds to Exposure Category F2 and F3. The recommended average total air contents in Table 3.1 for MODERATE and SEVERE exposure levels match target air content values in ACI 318.

Required average compressive strength f'_{cr} = 4200 psi (29 MPa)
Air content = 6 percent
Slump = 4 in. (100 mm)
Coarse aggregate available has dry-rodded density of 98 lb/ft³ (1570 kg/m³)
Fine aggregate has a fineness modulus of 2.60
Note: The average compressive strength, f'_{cr}, is the one required by ACI 301, Section 4.2 when f'_c has been specified as 3000 psi (21 MPa) and the producer has no strength data with the specified materials:

$$f'_{cr} = f'_c + 1200 \text{ psi} = 4200 \text{ psi (in.-lb units)}$$

$$f'_{cr} = f'_c + 8.3 \text{ MPa} = 29.3 \text{ MPa (SI units)}$$

Step 1: Choose slump—In this example, it is given as 4 in. (100 mm). If it were not given, a Table 6.3.1 from ACI 211.1 could be used to select an appropriate slump.

Step 2: Choose nominal maximum size of aggregate—Because the slab is 5 in. (125 mm) thick, the nominal maximum aggregate can be as large as 1/3 the thickness or 1-2/3 in. (42 mm). Use a standard maximum size 1-1/2 in. (37.5 mm) for the coarse aggregate in this unreinforced slab. For reinforced concrete, nominal maximum aggregate size is 3/4 of the smallest clearance between and around reinforcing bars.

Step 3: Estimate mixing water and air content—In the absence of data based on particular aggregates available, estimate the water content from Table 3.1. This table is condensed from ACI 211.1.

Because the slump was specified as 4 in. (100 mm), enter Table 3.1 on the second line of data, where the slump is shown as 3 to 4 in. (75 to 100 mm). Move to the column for 1-1/2 in. (37.5 mm) aggregate selected in Step 2. The table tells us how much water is needed per cubic yard (meter) of concrete:

275 lb (163 kg)

The air content was specified as 6 percent, and reading further down in the same column it is found that this is adequate for severe exposure of the concrete (as the minimum required total air content is 5.5 percent).

Step 4: Select the w/c or w/cm—Long experience and the accumulation of test information shows a relationship between w/cm and compressive strength of concrete. Refer to Table 3.2 for more information.

Because the code requires a concrete developed at 4200 psi (29 MPa), look for the w/cms for 4000 psi (28 MPa) and 5000 psi (35 MPa) air-entrained concretes in Table 3.2 and interpolate as:

$(0.48 - 0.40) \times (4200 \text{ psi} - 4000 \text{ psi})/(5000 \text{ psi} - 4000 \text{ psi})$
$= 0.016$ (in.-lb units)

$(0.48 - 0.40) \times (29.3 \text{ MPa} - 28 \text{ MPa})/(35 \text{ MPa} - 28 \text{ MPa})$
$= 0.015$ (SI units)

$0.48 - 0.02$ (rounded off) $= 0.46$

Use a w/cm of 0.46.

Note the w/cm of 0.48 for 4000 psi (28 MPa) concrete was reduced, because stronger concrete is needed for this slab.

Step 5: Calculate the cement content per cubic yard (meter) of concrete—Use the values for mass of water and w/cm from Steps 3 and 4.

Example (in.-lb units):
From Step 3, it is known:

mass of water per cubic yard of concrete = 275 lb

Table 3.2—Probable Minimum Average Compressive Strength of Concrete for Various Water Cementitious Materials Ratios (w/cm) (ACI 211.1)

Compressive strength at 28 days	w/cm by mass	
	Non-air-entrained concrete	Air-entrained concrete
6000 psi	0.41	-
5000 psi	0.48	0.40
4000 psi	0.57	0.48
3000 psi	0.68	0.59
2000 psi	0.82	0.74

Compressive strength at 28 days	w/cm by mass	
	Non-air-entrained concrete	Air-entrained concrete
40 MPa	0.42	-
35 MPa	0.47	0.39
30 MPa	0.54	0.45
25 MPa	0.61	0.52
20 MPa	0.69	0.60
15 MPa	0.80	0.71

From Step 4, it is known:

$$w/cm = \frac{\text{mass of water}}{\text{mass of cement}} = 0.46$$

Calculation for Step 5:

$$\text{mass of cement per cubic yard of concrete} = \frac{275 \text{ lb}}{0.46} = 598 \text{ lb}$$

Example (SI units):
From Step 3, it is known:

mass of water per cubic yard of concrete = 163 kg

From Step 4, it is known:

$$w/cm = \frac{\text{mass of water}}{\text{mass of cement}} = 0.46$$

Calculation for Step 5:

$$\text{mass of cement per cubic meter of concrete} = \frac{163 \text{ kg}}{0.46} = 354 \text{ kg}$$

Step 6: Estimate the coarse aggregate content—Again, the values for the trial mixture from ACI 211.1 are given as shown in Table 3.3. For equal workability, the volume of coarse aggregate in a unit volume of concrete depends only on its maximum nominal size and the fineness modulus of the fine aggregate to be used. Enter Table 3.3, the left

Table 3.3—Volume Of Coarse Aggregate Per Unit Volume of Concrete (From ACI 211.1)

Nominal maximum size of aggregate	Volume of dry-rodded coarse aggregate per unit volume of concrete for indicated fineness moduli of fine aggregate			
	2.40	2.60	2.80	3.00
3/8 in. (9.5 mm)	0.50	0.48	0.46	0.44
1/2 in. (12.5 mm)	0.59	0.57	0.55	0.53
3/4 in. (19 mm)	0.66	0.64	0.62	0.60
1 in. (25 mm)	0.71	0.69	0.67	0.65
1-1/2 in. (37.5 mm)	0.75	0.73	0.71	0.69
2 in. (50 mm)	0.78	0.76	0.74	0.72
3 in. (75 mm)	0.82	0.80	0.78	0.76
6 in. (150 mm)	0.87	0.85	0.83	0.81

column, down to the line for maximum aggregate size of 1-1/2 in. (37.5 mm). Move across to the column for fine aggregate with a fineness modulus of 2.60, and read 0.73 as the unit volume of coarse aggregate. Because this is a cubic-yard (cubic-meter) batch, multiply 0.73 by 27 ft^3/yd^3 (1 m^3) to get

19.71 ft^3 of coarse aggregate (per one cubic yard of concrete)

0.73 m^3 of coarse aggregate (per one cubic meter of concrete)

The table values are for dry rodded aggregate, so multiply coarse aggregate volume by the dry-rodded density of the aggregate to get mass of coarse aggregate (per cubic yard (meter) of concrete):

19.71 ft^3 × 98 lb/ft^3 = 1930 lb (in.-lb units)

0.73 m^3 × 1570 kg/m^3 = 1146 kg (SI units)

Step 7: Estimate the mass of fine aggregate needed—Following the weight method of ACI 211.1, simply add up the masses of cement, water, and coarse aggregate already determined.

275 lb + 598 lb + 1940 lb = 2813 lb (in.-lb units)

163 kg + 354 kg + 1146 kg = 1663 kg (SI units)

Then subtract the above value from the estimated mass of a cubic yard (meter) of concrete to find out how much fine aggregate is needed. Table 3.4 is used to obtain an estimated mass. Find on the line for 1-1/2 in. (37.5 mm) maximum aggregate to the column for air-entrained concrete. Read the estimated density as 3910 lb/yd^3 (2320 kg/m^3). Then, calculate the amount of fine aggregate per cubic yard (meter) of concrete:

3910 lb – 2813 lb = 1097 lb (in.-lb units)

2320 kg – 1663 kg = 657 kg (SI units)

SUMMARY: Now there are estimated quantities for a cubic yard (meter) batch, containing:
- Water: 275 lb (163 kg)
- Cement: 598 lb (354 kg)
- Fine aggregate: 1097 lb (657 kg)
- Coarse aggregate: 1940 lb (1146 kg)

Because entrained air was specified, include an air-entraining admixture, following the manufacturer's recommendations for the amount. This will probably be only a few ounces (grams), and will not affect the other values already determined. The entrained air does affect the volume of concrete, but this was allowed for in Table 3.4.

Once these seven steps are completed, the technician or mixture design specialist should make some adjustments by reducing the mixing water and increasing the amount of aggregate to account for free moisture on the aggregates. Then the amounts are scaled down to produce a small trial batch—for example, 2 ft^3 (0.06 m^3)—in the laboratory.

By testing the laboratory batches and making adjustments, of which usually several are needed, a mixture with satisfactory strength and workability is achieved. The quantities

Table 3.4—Estimated Density of Fresh Concrete (From ACI 211.1)

Nominal maximum size of aggregate	First estimate of concrete density, lb/yd^3 (kg/m^3) of concrete	
	Non-air-entrained concrete	Air-entrained concrete
3/8 in. (9.5 mm)	3840 (2278)	3710 (2201)
1/2 in. (12.5 mm)	3890 (2308)	3760 (2231)
3/4 in. (19 mm)	3960 (2349)	3840 (2278)
1 in. (25 mm)	4010 (2379)	3850 (2284)
1-1/2 in. (37.5 mm)	4070 (2415)	3910 (2320)
2 in. (50 mm)	4120 (2444)	3950 (2343)
3 in. (75 mm)	4200 (2492)	4040 (2397)
6 in. (150 mm)	4260 (2527)	4110 (2438)

CHAPTER 3—MIXTURE PROPORTIONING

can then be scaled up to a full-size field batch (refer to ACI 211.1) if you want to study the alternative absolute volume method for determining the trial batch amounts. It is more precise, but also more complex to perform.

Concrete for the small job

Although ready mixed concrete is widely used for most work, when you need less than 1 yd^3 (m^3) of concrete, you may not always find a ready mixed concrete producer to deliver it. Small batches may have to be mixed at the jobsite (Fig. 3.4). If mixture proportions or specifications are not available, you can use recommendations provided by the PCA in its book, "Concrete for Small Jobs"[3]. This book provides proportions by both mass and volume and gives instructions for making adjustments for local conditions and for mixing the concrete properly. Table 3.5 shows a sample of these recommendations for proportioning by volume when 3/4 in. (19 mm) and 1-1/2 in. (37.5 mm) maximum size aggregates are used.

Prepackaged dry concrete mixtures available from builders' supply and hardware stores are also appropriate for

Fig. 3.4—A portable drum mixer can be used on-site to mix smaller quantities of concrete (photo courtesy of PCA)

Table 3.5—Parts by Volume* for Making a Small Trial Batch of Concrete ("Concrete for Small Jobs"[3])

Maximum size coarse aggregate	Air-entrained concrete				Concrete without air			
	Cement	Wet sand	Wet coarse aggregate	Water	Cement	Wet sand	Wet coarse aggregate	Water
3/4 in. (19 mm)	1	2-1/4	2-1/2	1/2	1	2-1/2	2-1/2	1/2
1-1/2 in. (37.5 mm)	1	2-1/4	3	1/2	1	2-1/2	3	1/2

*The combined volume is about 2/3 of the sum of the bulk volumes of the ingredients.

jobs small enough for hand mixing. It is easy to choose packages to provide the concrete quantity required. Mixing directions and the correct amount of water to add are provided on the package. The bagged mixtures are most convenient and economical when only a few cubic feet (fraction of a cubic meter) are needed. For jobs up to about a cubic yard (meter), compare the cost of the prepackaged mixture with the cost of buying the ingredients separately. For jobs requiring a cubic yard (meter) or more, ready mixed concrete truck delivery is usually the most practical if available.

CHAPTER 4—BATCHING AND MIXING CONCRETE

Once the proportions for a concrete mixture have been determined as explained in Chapter 3, the quantities of ingredients for a given batch should be accurately measured and placed in the mixer. The process is referred to as *batching*. This chapter describes the batching process and describes how the concrete is mixed. ASTM C94/C94M, "Standard Specification for Ready-Mixed Concrete," is usually the controlling specification, unless the purchaser has customized a specification for a particular job.

Batching

Most specifications require that batching be done by mass rather than by volume because weighing gives the accurate measurement needed to assure concrete of uniform quality. However, quantities of water and liquid admixtures can be determined accurately by either volume or mass. Batching by volume is used for concrete mixed in a continuous mixer—see "Mobile batcher mixer (continuous mixer)" section—and sometimes for small jobs where there is no weighing equipment.

Batching can be done manually, semi-automatically, or automatically (Fig. 4.1). In a semi-automatic plant, the operator

Fig. 4.1—Typical control panel used for automatic batching in a ready mixed concrete plant. Some plants have computerized control of adjustments for batching their standard mixtures

uses push buttons or switches to open bins of material; gates close automatically when the proper amount of material has been released. In automatic batching, a single starter switch controls all the materials. ACI 117, "Specification for Tolerances for Concrete Construction and Materials and Commentary," requires batching accuracy for site-mixed concrete when quantities weighed are more than 30 percent of the full scale capacity, as follows:

Cementitious materials	± 1 percent
Aggregates	± 1 percent
Water	± 1 percent
Admixtures	± 3 percent

ASTM C94/C94M has similar requirements, with a few exceptions, for ready mixed concrete. These percentages are given as plus or minus (±), so that the accuracy of cement measurement for example could be 1 percent more or 1 percent less than the specified amount. Chemical admixtures, such as water-reducing admixtures and air-entraining admixtures are added to the mixture as solutions; the water mixed with them is considered part of the mixing water.

Mixing concrete

Concrete should be thoroughly mixed for the specified time until all the materials are evenly distributed. Mixers should be capable of discharging the concrete without harmful segregation, which would be without having large aggregate pieces clumped together. Where increased output is needed, use a larger mixer or additional mixers, instead of speeding up or overloading the equipment available. Note that ASTM C94/C94M limits truck-mixed concrete to 63 percent of the total volume of the drum.

If the concrete has been properly mixed, samples taken from different portions of a batch will have essentially the same unit weight (density), air content, slump, and coarse-aggregate content. ASTM C94/C94M gives maximum allowable differences in test results within a batch of ready mixed concrete, as shown in Table 4.1. The batch is considered satisfactory if it meets five of the six uniformity checks.

Stationary mixers: central or on site

Stationary mixers may mix concrete at the jobsite, or they may be used as central mixers in ready mixed concrete plants. The more common types are:
- Tilting drum mixer: Revolving drum discharges by tilting the axis of the drum.
- Non-tilting drum: Revolving drum is charged, mixes, and discharges with the axis of the drum horizontal.

Table 4.1—Permissible Differences Between First and Last Portions of a Batch of Concrete (ASTM C94/C94M)

Test	Maximum permissible difference
Mass per cubic foot (mass per meter) of concrete, calculated to an air-free basis	1 lb/ft^3 (16 kg/m^3)
Air content (volume percent)	1% (numerical difference in percent units)
Slump: 4 in. (100 mm) or less	1 in. (25 mm)
Slump: 4 to 6 in. (100 to 150 mm)	1.5 in. (40 mm)
Coarse aggregate content, portion retained on No. 4 (4.75 mm) sieve (percent by mass)	6% (numerical difference in percent units)
Mass per unit volume of air-free mortar	1.6% (numerical difference in percent units)
Average compressive strength, 7 days	7.5% (numerical difference in percent units)

Note: A satisfactory batch is required to meet at least five of the six limits.

- Vertical shaft mixer: Often called turbine or pan-type mixer. Mixes with rotating blades or paddles mounted on a vertical shaft in a stationary or rotating pan.

Stationary mixers can be equipped with loading skips and some have a swinging discharge chute. Many stationary mixers have timers that prevent discharge of the batch before the designated mixing time has elapsed.

For stationary mixers obtaining a preblending or ribboning effect as the stream of materials enter the mixer is essential. Stationary mixing usually requires at least 1 minute mixing time for mixers up to 1 yd^3 (0.76 m^3) capacity, plus 15 seconds for each additional 1 yd^3 (0.76 m^3) of capacity, but highway specifications permit less mixing time.

Timing starts after all the ingredients, except the last of the water, are in the mixer. Overmixing can drive out entrained air and increase fines by grinding of the aggregate particles. Limits are placed on maximum as well as minimum mixing time.

Except when water is heated, a 5 to 10 percent portion of the mixing water should be added before the other materials. Then 80 percent of the remaining water is added uniformly with the solid materials, leaving about 10 percent to be added later; this is sometimes referred to as "hold-back water." If the mixer is charged directly from batchers, all of the materials can be added simultaneously at rates that keep the charging time the same for all materials.

Chemical admixtures are usually in solution, and are generally added to the mixing water at the time of batching, no later than 1 minute after addition of the water to the cement has been completed, before the start of the last 3/4 of the mixing cycle, or whichever occurs first. Finely divided SCMs are added with the dry ingredients. High-range water-reducing admixtures and fibers are sometimes added at the jobsite. Sufficient mixing time should be given after a site addition to ensure adequate dispersement of the admixture or fibers. The mixing time for fibers in repulpable bags should be carefully determined as the type of aggregate, which can be either angular or rounded, can impact the mixing action.

Ready mixed concrete

Ready mixed concrete is manufactured for delivery to a purchaser while it is plastic and unhardened. It may be central-mixed, shrink-mixed, or truck-mixed.

Central-mixed concrete (Fig. 4.2) is discharged for delivery into either a truck agitator, a truck mixer operating at agitating speed, or in a special non-agitating truck. Shrink-mixed concrete is a ready mix product that is partially mixed in the central stationary mixer, then discharged into a truck mixer that completes the mixing according to specifications. For truck-mixed concrete, all of the mixing is done in the truck that delivers the concrete. ASTM C94/C94M covers all three kinds of ready mixed concrete. Anyone who verifies that the number of revolutions at mixing speed is within the prescribed limits should know which kind of mixing is used.

Proper loading of truck mixers can prevent packed sand and cementitious materials in the drum head during charging or loading. Adding about 10 percent of the coarse aggregate

Fig. 4.2—Central mixing in stationary mixer of the tilting drum type (photo courtesy of PCA)

Fig. 4.3—Ready mixed concrete trucks deliver centrally mixed concrete or mixture ingredients that have been batched directly into the truck

and water into the drum ahead of the sand and cementitious materials is helpful. Generally about 1/4 to 1/3 of the water should be added into the discharge end of the drum after all other ingredients have been loaded. Liquid admixtures should be added with the water or on damp sand. Powdered admixtures are ribboned into the mixer with other dry ingredients.

Truck mixers have two operating speeds: 1) *agitating speed*—usually about 2 to 6 revolutions per minute; and 2) *mixing speed*—about 6 to 18 revolutions per minute.

When a truck mixer is used for complete mixing, 70 to 100 revolutions of the drum or blades at the rate of rotation designated by the manufacturer as "mixing speed" are usually required to produce the specified uniformity of concrete. Mixing at high speeds for long periods, an hour or more, can result in concrete strength loss, temperature rise, excessive loss of entrained air, and accelerated slump loss.

If there is a time lag between mixing and discharge, the drum speed is reduced to the agitating speed or stopped. Then prior to discharging, the mixer is again operated at mixing speed for about 30 revolutions to improve uniformity.

ASTM C94/C94M requires that concrete be delivered and discharged within 1.5 hours or before the purchaser specified limit on drum revolutions after introduction of water to the cement and aggregates or the cement to the aggregates. The purchaser may waive these limitations if the slump is such that the concrete can be placed without the addition of water.

Mobile batcher mixer

Mobile batcher mixer (continuous mixer) trucks (Fig. 4.4) are combination materials transporters and mixers. This type of mixer is often called a volumetric mixer. They batch by volume and continuously mix concrete as the dry materials, water, and admixtures are continuously fed into the auger mixer. ASTM C685/C685M, "Standard Specification for Concrete Made by Volumetric Batching and Continuous Mixing," covers this operation. Concrete can be proportioned and mixed at the jobsite by one operator, on an as-needed basis. The concrete mixture is easily adjusted for project placement and weather conditions.

High-energy mixers

High-energy mixers first blend cementitious materials and water into a slurry with high-speed rotating blades. The slurry is then added to aggregates and mixed with conventional mixing equipment to produce a uniform concrete mixture.

High-energy mixing combines water more completely with cement particles, resulting in more complete cement hydration. This results in more efficient use of cement, higher strength, and improvements in several other concrete properties over those developed by conventional mixing alone.

Remixing concrete

When concrete arrives at the jobsite too stiff for placing, a small amount of water can be added provided the maximum permissible *w/cm* and the maximum specified slump are not exceeded. After adding water, ASTM C94/C94M requires an additional 30 revolutions or more at mixing speed for the truck mixer. Remixing of ready mixed concrete should be done within the ASTM C94/C94M limits on maximum mixing and agitating time or limits on revolutions of the mixer drum.

Do not indiscriminately add water to make the concrete more fluid because this lowers the quality of the concrete. Adding the 20 lb (about 2.5 gallons) of water per 1 yd^3 (12 kg (or about 12 L) of water per 1 m^3) that may be needed to increase slump by 2 in. (50 mm) can cause a loss of 200 psi (1.4 MPa) or more in compressive strength. Added water can also increase finishing time, decrease the durability of concrete, and will definitely increase shrinkage. Remixed

Fig. 4.4—Bins and tanks to store materials are needed by the mobile batcher mixer. Dry ingredients, water, and admixtures are measured by volume and mixed continuously

concrete can be expected to harden rapidly and cold joints may result when concrete is placed beside or above remixed concrete.

Maintenance of mixing equipment

If the blades of the mixer (Fig. 4.5) become worn or coated with hardened concrete, the mixing action is less efficient. Badly worn blades should be replaced and hardened concrete should be removed periodically, preferably after each day of concrete production. It is advisable to require or obtain certification of batch plants and delivery trucks to verify that they are in good working order. Certification can typically be obtained from the state Department of Transportation or the National Ready Mixed Concrete Association (NRMCA).

Fig. 4.5—Mixer blades should be cleaned after use to prevent build-up of hardened concrete

CHAPTER 5—HANDLING, PLACING, AND CONSOLIDATING CONCRETE

Good advance planning is needed to choose the best method of handling the concrete, once it has been mixed and delivered as described in Chapter 4. Schedules and methods should be set up carefully to avoid:
- Delays
- Early stiffening and drying
- Segregation

Delays will affect productivity and profitability of the job, as well as the quality of finished work. Stiffening and drying may become a problem when the concrete is held more than 1.5 hours before placement and consolidation. Delays are even more critical in hot, dry weather, when accelerating admixtures are used, or a heated mixture is used, or when the concrete has a high cement content or low *w/cm* ratio.

Preconstruction meetings between all the parties involved (ready mixed concrete supplier, concrete contractor, and project engineers or owners) are recommended to clearly define the responsibilities of each party. The National Ready Mixed Concrete Association (NRMCA) and the American Society of Concrete Contractors (ASCC) have published a suggested list of items to be covered at a preconstruction meeting. The NRMCA/ASCC "Checklist for the Concrete Pre-Construction Conference"[4] can also be found in the ACI/ASCC Contractor's Guide to Quality Concrete Construction.[5]

Segregation is the separation of some of the coarse aggregate from the sand-cement mortar. This leaves part of the batch having too little coarse aggregate and the rest of it having too much. Where there is too little coarse aggregate the concrete is likely to shrink more and crack and have poor resistance to abrasion. Where there is too much coarse aggregate, the concrete is too harsh for full consolidation and finishing and may develop voids called *honeycomb*. Avoiding segregation is a major concern when selecting the equipment to transport and handle the concrete.

When practical, placing concrete directly from the truck mixer is usually the most convenient and economical. When the point of placement is beyond the reach of the truck mixer's chute, other methods are used. The most common of these are chute extensions, buggies, belt conveyors, crane and bucket placement, and pumping. Drop chutes, tremies, pneumatic guns, shotcrete equipment, and screw spreaders are also used for special applications.

The concrete mixture specification may affect the choice of method. Slump, type or size of aggregate, or use of lightweight or heavyweight concretes may favor one method of placement over another. On larger jobs, it may be the other way around. When the job schedule dictates use of a certain delivery method—such as pumping, for example—the mixture may be adjusted through use of admixtures or changes in the proportions to improve its pumpability.

Handling and placing methods

This section describes briefly the principal methods of handling and placing concrete. For detailed information refer to ACI 304R, "Guide for Measuring, Mixing, Transporting, and Placing Concrete."

Depositing concrete from the truck mixer

Most ready mixed concrete specifications permit only one addition of water, and only enough to increase the slump to be within the range specified without exceeding the specified w/cm. After the water is added, the mixer drum is turned 30 revolutions at mixing speed. The truck chute is then wetted to facilitate the flow of concrete. Do not add water after the load of concrete has been partially discharged.

Direct deposit from the truck mixer (Fig. 5.1) works best when the forms are at or below grade. The angle at which the chute is held during placement is important. Sometimes chute extensions are used, adding as much as 15 ft (4.6 m) to the reach of the truck. When the chute slope becomes too flat, however, concrete with less than about 4 in. (100 mm) slump may not flow well down the chute.

When performing slab-on-ground construction, trucks should not be driven inside the forms unless the subgrade can support the trucks without rutting or if the rutting is corrected during placement by using equipment such as a laser box grader. Where placing conditions do not allow the concrete to flow down the chute at the specified slump, a suitable solution may be to add a high-range, water-reducing admixture, if

Fig. 5.1—Where the site provides solid support for the ready mix truck close to the forms, direct chute discharge from the truck is the simplest way to place concrete (photo courtesy of PCA)

the specifications allow it, so long as the addition does not increase shrinkage and the addition remains consistent so as not to result in variable setting and bleeding behavior. Where access and turnaround are restricted, front-end discharge truck mixers (Fig. 5.2) can be useful.

Wheelbarrows and buggies

Common wheelbarrows (Fig. 5.3) and Georgia buggies, which are manually operated buggies with two large wheels, are sometimes used for small jobs, but powered concrete buggies (Fig. 5.4) are much more common. Wheelbarrows and buggies are useful for short hauls on all types of construction, especially where access to the work area is limited. They are valuable on jobs where placing conditions are constantly changing. However, they are slow. Discharging concrete into buggies and depositing concrete in place from the buggy requires more labor than other methods, and may be at a disadvantage where there is high-volume demand.

Walkways or runways for buggies should always be supported from grade or from the deck form, not by the reinforcing steel. They should be smooth to prevent jarring the concrete. Mixtures with slumps of 4 in. (100 mm) or more will segregate when subjected to excessive jarring.

Fig. 5.2—Front-end discharge truck mixers give the driver better mobility and control when discharging concrete directly into forms (photo courtesy of PCA)

Fig. 5.3—Wheelbarrows, preferably with pneumatic tires, are useful on small jobs where the wheeling distance is short and the quantity of concrete is small. Proper access to the top of the form should be provided (photo courtesy of PCA)

Fig. 5.4—The versatile power buggy, good for short hauls may be a riding model like this one, or a walk-behind model (photo courtesy of PCA)

Runways should be laid out in a circular or loop path so that traffic can move continuously and safely. Plan ahead for adjustments as the placement progresses so that traffic flow can continue uninterrupted.

Belt conveyors

Rigidly supported conveyor belts are widely used to maneuver concrete horizontally or to a higher or lower level. Concrete is deposited through hoppers onto the belt in a continuous ribbon. With adjustable reach and variable speed both forward and reverse, they make it possible to deposit large volumes of concrete rapidly in relatively inaccessible areas. Placement capacities for a single conveyor from 150 to 700 yd^3 (110 to 540 m^3) per hour are not unusual.

Some conveyors are mounted on truck mixers, some travel to the job on their own truck (Fig. 5.5). Simpler ones may be towed to the site or hauled along with other equipment. When using conveyors, segregation at the point of discharge is avoided by discharging through a suitable drop chute. Concrete slumps of 2 to 4 in. (50 to 100 mm) are generally recommended when conveyors are used because segregation may occur; the sand-cement mortar will separate from the aggregate if mixtures are too wet. Air-entrainment is helpful for mixtures with a tendency to segregate.

In hot, dry, windy weather, any long line of chutes or belts should be covered to prevent drying of the concrete and excessive slump loss. Conveyors also have to be maintained daily during use. The hoppers, belts and scrapers have to be cleaned, and before use the following

Fig. 5.5—This conveyor came to the site mounted on its own truck. The long, adjustable reach carries concrete above site obstructions, and the drop chute at the end helps prevent segregation (photo courtesy of PCA)

day the system should be run briefly to confirm it is operating properly before concrete arrives. When chutes or belts are flushed with water for cleaning, the water and diluted concrete should not be allowed to drain into the forms or onto freshly placed concrete. Prewetting chutes or belts helps concrete flow into place more easily and facilitates cleanup.

Buckets and hoppers

Bottom dump buckets and hoppers permit placement of the lowest possible slump concrete. They should have side slopes not less than 60 degrees, with wide, freeworking and tight-closing discharge gates. The buckets of concrete are delivered in several different ways. Bucket and crane placement is used on many large projects. On very large work, such as dams, cableways may move the concrete bucket or hopper to the point of deposit. On occasion helicopters have even been used to lift and deliver concrete buckets.

Placing concrete by bucket or hopper is extremely flexible because of the range of vertical and horizontal distances that can be reached by a crane from one point of deposit. Conventional cranes stand away from the building the distance needed to accommodate the boom. Tower cranes stand immediately adjacent to or within the structure. Self-climbing cranes rise with the structure. Crane operation should be planned considering the maximum reach with a loaded bucket, the rate of swing, and rebound upon dumping. Allow for the worst case conditions unless positive measures are taken to prevent them.

The bucket's discharge gate should be readily controllable to limit the rate of placement so that concrete can be spread as it is being discharged. As with other methods, concrete should not be deposited in a pile and spread with a vibrator. This can cause segregation and nonuniform concrete.

Pumping concrete

Concrete pumps can convey concrete directly from a central discharge point at the jobsite directly to the formwork (Fig. 5.6), reaching locations difficult to access otherwise. The same pump can move concrete both vertically and horizontally through pump lines that take up little space, and can be moved or extended readily.

Truck-mounted pumps can be delivered to the job when needed, either rented by the contractor or supplied by a pumping subcontractor who provides the service. Stationary and self-climbing pump booms, which are also called placing booms, are used to provide continuous concrete for tall buildings.

Concrete mixtures for pumping should be plastic, not harsh, and the proportions are specially selected for pumpability as well as for strength and other specified properties. Air-entrainment, chemical admixtures, and the addition of fly ash can improve pumpability and reduce the danger of separating the cement-sand mortar from the coarse aggregate under pressure. Rounded aggregates (gravel) will pump more easily than crushed aggregates.

To lubricate the pump line, start the operation by pumping a properly proportioned mortar or a batch of the regular concrete with the coarse aggregate omitted. Once concrete flow through the pipeline is established, lubrication is maintained as long as pumping continues. The line coating mixture should be wasted and not used in the concrete placement.

Fig. 5.6—Concrete pumps can deliver concrete from a central discharge point to hard-to-reach locations on a congested project site

The pump and pipe should be cleaned thoroughly upon the completion of pumping. Because concrete pumping is a continuous operation, proper scheduling of ready mix trucks is critical. A long delay may require having to break down the line, clean out, and start again.

When lightweight concrete is pumped, it is necessary to presoak the lightweight aggregate before batching. Otherwise, under pressure from the pump the aggregate may absorb enough water from the cement paste to make the mixture too stiff for proper pumping. There are several ways to presoak lightweight aggregates; some vacuum and thermal processes require as little as 45 minutes. Continuous sprinkling of aggregate piles may take up to 3 days for the aggregate to reach saturation.

Because slump and air loss may occur during pumping, some specifications may call for tests of concrete taken at the point of discharge from the pump line. Often, once a consistent slump or air loss is identified and the properties at the end of the pump line (point of placement) can be predicted, only testing the concrete at the pump (point of delivery) can be resumed.

Pneumatic or air gun placing

Another method of conveying concrete through pipelines is pneumatic placing, commonly used for tunnel work. Air pressure is used to force concrete through the pipes, with discharge lines either horizontal or inclined upward from the placer. Careful control is needed to get uniform in-place concrete. A slow, nonviolent discharge is required until the end of the pipe can be immersed in the concrete. Because loss of slump between the mixer and the forms can be 2 to 3 in. (50 to 75 mm), it is necessary to start out with a wetter mixture than that desired in the forms.

Fig. 5.7—Steel bucket, moved by crane for building construction, has circular lever for operating discharge gate. Bucket capacities range from 3/8 to 15 yd³ (0.3 to 11.5 m³)

Shotcrete

Shotcrete is mortar or concrete pneumatically projected at high velocity onto a surface. Although there are many different pieces of proprietary equipment for shotcreting, shotcrete for normal construction requirements is produced by either the dry-mix or wet-mix process.

In the *dry-mix process*, the cementitious materials and aggregates are mixed dry and

Fig. 5.8—With hydraulically operated booms spanning 150 ft (45 m) or more, concrete pumps can reach over rough terrain and intermediate structures. Truck-mounted pumps move readily from one part of the site to another

Fig. 5.9—Shotcrete can be built up gradually in layers, encasing reinforcing steel while building the desired thickness of section. Generally formwork is required for only one side (photo courtesy of American Shotcrete Association)

carried by compressed air through the delivery hose to a nozzle body where water is mixed under pressure with the other ingredients; the mixture is jetted from the nozzle at high velocity onto the surface to be shotcreted.

In the *wet-mix process*, all materials except possibly an accelerator, are thoroughly mixed, then conveyed through the delivery hose to the nozzle. Additional air is injected at the nozzle, and usually an accelerator is added. Then the mixture is jetted at high velocity onto the surface to be shotcreted.

Shotcrete is used where concrete is being placed in difficult locations and where large areas of thin cross section are needed. It is ideal for repair work and curved and free-form shapes, generally requiring only a one-sided form or sometimes no formwork at all, for example a swimming pool. The quality of the work depends greatly on having a skilled operator or nozzleman at the nozzle. This is such an important job that certain projects require certification of nozzle operators to assure quality work. ACI offers certification programs for Shotcrete Nozzleman Operators to demonstrate the proper application of shotcrete.

Other placing equipment and methods

Open discharge chutes from ready mixed concrete trucks, concrete buckets, and conveyors typically are sloping. To avoid segregation, a short downpipe (Fig. 5.10) at the end of the sloping chute can direct the concrete into a vertical fall. Concrete is frequently deposited into the forms through drop chutes or elephant trunks (Fig. 5.11) to avoid striking reinforcement or sides of the forms. If the entire placement can be finished before any of the concrete splashed on forms or reinforcement dries, and if concrete can be deposited between vertical curtains of reinforcement, drop chutes may not be needed.

A tremie pipe is used for placing concrete under water, funneling concrete down through the water into the required location. The discharge end of the tremie is buried in fresh concrete, keeping a seal between the water and the concrete mass, letting the concrete "mushroom."

Screw spreaders are usually used as part of a paving train of equipment. With a screw spreader the concrete discharged from a truck or bucket can be quickly spread over a wide, flat area to a uniform depth. The screw spreader leaves the concrete fairly well compacted but final consolidation by vibration is required.

Where concrete is to be placed under water, structures are heavily reinforced, or concrete with low volume change is required, preplaced aggregate concrete may be used. In this method of placement, the forms are first filled with clean coarse aggregate, then structural quality grout is pumped through injection pipes to completely fill all of the voids.

CORRECT

The above arrangement prevents separation, no matter how short the chute, whether concrete is being discharged into hoppers, buckets, cars, trucks, or forms.

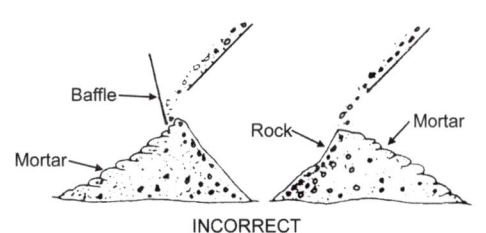

INCORRECT

Improper or lack of control at the end of any concrete chute, no matter how short. Usually a baffle merely changes direction of separation.

CONTROL OF SEPARATION AT THE END OF CONCRETE CHUTES

This applies to sloping discharges from mixers, truck mixers, etc., as well as to longer chutes, but not when concrete is discharged into another chute or onto a conveyor belt.

Fig. 5.10—Control of separation at the end of sloping concrete chutes. A short downpipe is suggested to prevent segregation and direct the concrete into a vertical fall (from ACI 304R)

Depositing the concrete

All equipment used to place concrete should be kept clean and in good working condition. Standby equipment should be available in the event of a breakdown. Concrete should be deposited continuously as near as possible to its final position. In slab construction, concrete should not be dumped in separate piles and then leveled and worked together; nor should the concrete be deposited in large piles and moved horizontally into final position. Such practices result in

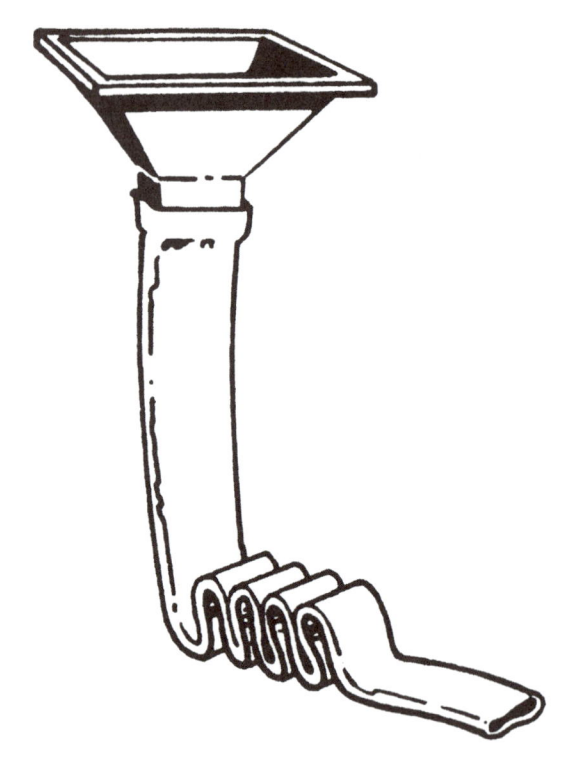

Fig. 5.11—Flexible drop chutes, or elephant trunks, are made of plastic or canvas in lengths up to 50 ft (15 m). They attach to hoppers and buckets and help prevent segregation and splattering of concrete on forms and reinforcement (CCS-2)

Fig. 5.12—Cross section of concrete that was poorly consolidated (photo courtesy of CTLGroup)

segregation because mortar tends to flow ahead of coarser material.

In deeper forms, concrete should be placed in uniform horizontal layers up to 20 in. (500 mm) thick, depending on the type of work. Each layer is thoroughly consolidated before the next is placed. To avoid flow lines and cold joints, the rate of placement should be fast enough that one layer of concrete has not set when the new layer is placed above it.

Consolidation

Consolidation, that is sometimes incorrectly referred to as compaction, is a process that eliminates most voids in the concrete mixture and brings the solid particles closer together. During consolidation, concrete should be worked well around reinforcement and embedded fixtures and into corners of the forms. Vibration, spading, and other manual and mechanical ways are used to achieve consolidation. The method chosen depends on the consistency of the mixture and the complexity of formwork and reinforcement. A mixture that can be consolidated easily with hand tools should not be mechanically vibrated, because it is likely to segregate.

Hand methods

Workable, flowing concrete can be consolidated by rodding—that is, thrusting a tamping rod repeatedly into the concrete. The rod should be long enough to penetrate the full depth of the layer being placed and thin enough to pass between the reinforcing steel and the forms.

Supplementary spading by hand can improve the visual surface quality of concrete that has been mechanically vibrated. A flat spade-like tool repeatedly inserted and withdrawn adjacent to the form forces larger coarse aggregates away from the form and reduces the number of air-bubble voids on the concrete surface.

Fig. 5.13—Internal vibrator action in low-slump concrete makes it behave temporarily as a liquid, settling into corners of the forms and eliminating air voids (photo courtesy of PCA)

For dry hand-tamped concrete, the surfaces are rammed with heavy flat-faced tools until a film of mortar or paste appears at the surface, indicating that the voids within have been filled.

Mechanical vibration

Mechanical vibration is the most widely used method for consolidating concrete. It consists of subjecting the concrete to rapid vibratory impulses that cause the concrete to behave temporarily like a liquid and settle in the forms. During vibration the large entrapped air voids rise more easily to the surface.

Vibrators for cast-in-place concrete are of three principal types:
- Internal vibrators
- Form vibrators
- Surface vibrators

Vibrators are usually characterized by their frequency of vibration, expressed as the number of vibrations per minute (VPM), and by the amplitude of vibration, which is the deviation in in. (mm) from the point of rest. ACI 309R, "Guide for Consolidation of Concrete," recommends which frequency and amplitude to use for different types of concrete.

Fig. 5.14—This flexible-shaft internal vibrator, driven by a gasoline engine, is usually carried on a back pack (photo courtesy of Minnich Manufacturing)

Internal vibration

Internal vibration is best suited for ordinary construction, provided the section is large enough for the vibrator to be used effectively. In inaccessible areas or places congested with reinforcement, other methods may supplement internal vibration. Internal or immersion-type vibrators are often called "spud" or "poker vibrators." Driven by electric motors, gasoline engines, or by hydraulic or pneumatic power, they are commonly used to consolidate concrete in walls, columns, beams, and slabs. The flexible-shaft vibrator, consisting of a vibrating head connected to the drive motor by a flexible shaft, is probably the most widely used. It and other common types of internal vibrators are described in ACI 309R.

Proper use of internal vibrators is important for best results. Vibrators should not be used to move concrete horizontally because this causes segregation. Whenever possible, the vibrator should be lowered vertically to the full depth of the layer being placed, and at least 6 in. (150 mm) into the preceding layer. It should be manipulated up and down for 5 to 15 seconds to knit the two layers together. Then the vibrator should be withdrawn gradually with a series of up and down motions. It is operated continuously while being withdrawn so that no holes are left in the stiff concrete. The vibrator is inserted at regular intervals, spaced about 1.5 times the radius of vibrator action so that the area visibly affected by the vibrator overlaps the adjacent just-vibrated area (Fig. 5.15).

If there is any doubt about the adequacy of consolidation, vibrate more. There is little danger of overvibrating a properly proportioned mixture.

In working on slabs, the vibrator should be sloped toward the horizontal far enough to operate in a fully embedded

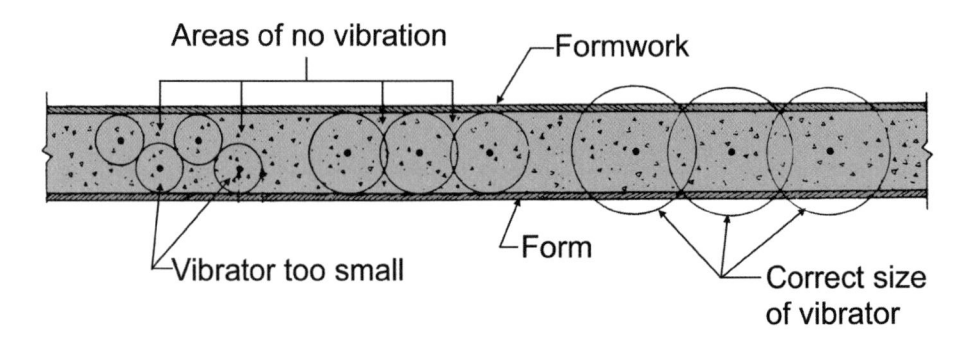

Fig. 5.15—Vibrator should be inserted at regular intervals spaced about 1.5 times the radius of the vibrator's action. In this drawing of a wall form, the black dots indicate vibrator insertion points and the circle shows radius of vibrator action. The less powerful vibrator on the left requires many more insertions to fully consolidate the concrete

position. For slabs-on-ground, the vibrator should not make contact with the subgrade.

Form vibration

Form vibrators are external vibrators attached to the outside of the form or mold. They vibrate the form or mold, which transmits the vibration to the concrete. Form vibrators may be either rotary or reciprocating type.

Form vibration is used mostly for precast concrete, but it is suitable for thin sections of cast-in-place concrete and is a useful supplement to internal vibration where it is difficult or impossible to insert an internal vibrator due to conditions such as congested reinforcement. Form vibrators are also used to reduce surface air voids or bug holes on formed surfaces. Form vibration can result in pressures substantially higher than normal, and particular consideration should be given to formwork design.

For vertical forms, high amplitude and low frequency of vibration are preferred for stiffer concretes. Low amplitude and high frequency work better with more plastic mixtures (ACI 309R).

Surface vibration

Surface vibrators, typically called vibrating screeds (Fig. 5.16), are used mainly in slab construction and are applied to the top surface. These tools consolidate concrete from the top down by maintaining a head of concrete in front of them. Their leveling effect helps in finishing the slab.

There are four types of vibratory screeds: 1) hand-operated wet screeds; 2) roller screeds; 3) truss screeds; and 4) laser screeds. All have a vibrating blade that consolidates the concrete as it is struck off to the required elevation. The hand-operated screeds work as a wet screed, using pins and side forms to help define the desired elevation. Truss or roller screeds are supported by the side forms or rails to finish to the required elevation. For structurally reinforced slabs, internal vibration may also be required to properly consolidate the concrete.

Laser screeds are large high-tech machines that strike off and consolidate the concrete using a laser beam to establish the proper elevation. Laser screeds allow contractors to place large industrial floors at high rates of placement and to achieve high tolerances for floor flatness. The screed head ranges from 12 to 20 ft (3 to 6 m) in width and is attached to a boom that can extend out to 20 ft (6 m). Laser receivers on each end of the screed head maintain proper elevation. The screed head includes an auger to place the concrete and a vibrator to consolidate it.

Roller compacted concrete

Roller compacted concrete (RCC) is a concrete of zero slump consistency that is used in dams and pavement construction. It is transported, placed, and compacted in horizontal layers using equipment that is similar to that used for highway construction and earth and rock-fill work. Consolidation is achieved using smooth-drum vibratory rollers weighing 1200 to 3000 lb/ft (1800 to 4500 kg/m). This technique produces a dense, well-compacted concrete.

Benefits of consolidation

A mass of freshly placed concrete is usually honeycombed with pockets of entrapped air. If the concrete is allowed to harden in this condition, it is nonuniform, weak, porous, and poorly bonded to the reinforcement. The mixture should be densified by the consolidation methods just described if it is to develop the strength, durability and other properties normally expected of concrete. Proper consolidation is also necessary to obtain the uniform surfaces demanded

Fig. 5.16—Vibrating power screed levels and consolidates slab-on-ground concrete. This one is powered to move forward automatically as the crew works. Smaller vibrating screeds may be hand-drawn

CHAPTER 5—HANDLING, PLACING, AND CONSOLIDATING CONCRETE 79

for architectural concrete. The most serious imperfections resulting from ineffective vibration are:
- Honeycomb
- Excessive bug holes
- Placement (lift) lines

Because each of these imperfections can be influenced by other mixing and placing conditions, poor vibration is not always to take full blame. For example, if concrete is badly segregated by improper handling and improper deposit in the forms, vibration may improve but will not remedy the entire problem.

Undervibration is far more common than overvibration in normalweight concrete. Thus, if there is any doubt as to the adequacy of the consolidation, it should be resolved by additional vibration.

Honeycombs (Fig. 5.17), also referred to as rock pockets, occur when the cement paste does not fill all the spaces between aggregate particles. When it shows on an exposed concrete surface, removal and repair may be required.

Bug holes (surface air voids) on formed concrete surfaces, as shown in Fig. 5.18, are regarded by some as a natural feature of concrete. But a problem occurs when they are too large, too numerous, or both. Attention to form release agents and the composition of the mixture, as well as to placement and consolidation procedures, will help reduce the bug holes. Spading near the concrete surface is a remedy already mentioned.

Placement or *lift lines* are dark lines on formed surfaces at the boundary between adjacent batches of concrete. Usually they indicate that the vibrator was not lowered far enough to penetrate the layer below the one being vibrated, or that too much time passed between placement of successive layers of concrete, and the concrete in place was already beginning to set.

Self-consolidating concrete

Self-consolidating concrete (SCC) is specifically proportioned to flow great distances without segregation. Because of its special characteristics, placing methods for SCC are much different than for normal concrete. For example, SCC is usually deposited in one location and allowed to flow throughout the formwork (Fig. 5.19). There is also no need for mechanical consolidation. Because of its high fluidity, SCC may produce much higher pressures on the formwork.

Fig. 5.17—Honeycombs or rock pockets are places where the cement paste does not completely fill the space between aggregate pieces. Poor vibration sometimes causes honeycomb (photo courtesy of PCA)

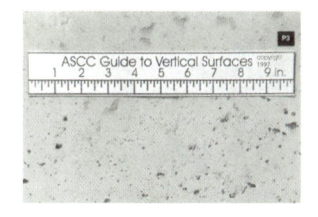

Fig. 5.18—Proper vibration procedures and adequate vibrator power help reduce bugholes like these. However, the mixture design, form release agent, and placing procedures also contribute to the prevalence of bug holes (1 in. = 25.4 mm) (photo courtesy of American Society of Concrete Contractors)

Fig. 5.19—Self-consolidating concrete (SCC) being placed. Note that the material flows throughout the formwork

CHAPTER 6—CURING AND PROTECTION

Curing is maintaining satisfactory temperature and moisture content in concrete long enough for the concrete to develop the potential properties. The potential strength and durability is fully developed only if the concrete is cured. Good curing usually requires keeping the concrete moist and above 50°F (10°C) until desired properties of the concrete are achieved. Improper curing can cut the strength of even the best concrete by as much as 50 percent (Fig. 6.1). Most of the time, it is necessary to start specified curing measures as early as possible. An early start of curing is critical for high-strength concretes, particularly those containing silica fume.

Remember that curing depends on three factors:
1. Moisture
2. Temperature
3. Time

Each of these factors are discussed briefly in this chapter. In addition, the special protection needed by concrete in hot weather, cold weather, and wind are presented.

Although the curing needs of concrete placed during extremely cold or hot weather remain the same as during more moderate temperatures, these conditions require more effort by workers. Curing and protection techniques for both types of weather extremes should be planned well ahead of time. For more information, refer to ACI 305R, "Guide to Hot Weather Concreting," ACI 306R, "Guide to Cold Weather Concreting," and ACI 308R, "Guide to Curing Concrete."

Rain damage

If rain falls on freshly placed concrete, the water may erode the surface of fresh concrete and dilute the cement paste at and near the surface. Preparations, therefore, should be made in advance for protec-

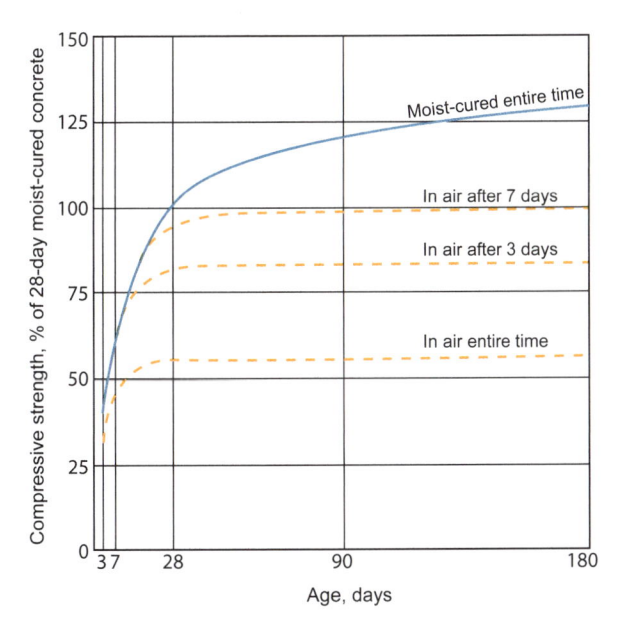

Fig. 6.1—Concrete dried in air may gain only half the strength of the same concrete cured moist for 7 days. Concrete cured moist for 180 days continues to gain strength throughout the curing period (from ACI 306R, Fig. 8.3)

Fig. 6.2—Rain can erode the surface of freshly placed concrete. Protect fresh concrete from rain using a material to cover the surface (photo courtesy of Michael Ayers)

Fig. 6.3—Rain damage on concrete surface (photo courtesy of Michael Ayers)

tion from rain when work must continue under such conditions (Fig. 6.2). Finishing unformed concrete cannot be worked on in the rain, and the rain will quickly damage new work (Fig. 6.3). Either shelter the work completely from the rain, or stop work until the rain is over.

Curing time and temperature

Warm concrete sets faster than cold concrete, and during the first few days gains strength faster than cold concrete. Where high strengths are needed quickly, concrete may be heated by steam or other means. At temperatures just above freezing, fresh concrete hardens very slowly and will freeze at concrete temperatures below 32°F (0°C) in cold weather. When concrete temperature falls below 40°F (4°C), cement hydration will stop. Thus, when air temperatures fall below 40°F (4°C), concrete should be protected for the first day or two. Cold weather precautions are discussed later in the chapter.

If concrete surfaces are allowed to dry before the concrete sets, or are allowed to be alternately wet and dry, cracks may appear on the surface. To prevent such cracks, evaporation during finishing and for several days thereafter should be reduced. All materials and equipment needed to prevent early drying and for curing should be available and ready for use before the concrete arrives. Curing of flatwork should begin immediately after placing and finishing of concrete have been completed.

When the temperature of the air is above 40°F (4°C), specifications frequently call for a minimum of 7 days of curing, or at least long enough for the concrete to attain 70 percent of its specified compressive or flexural strength. Formwork left in place offers some measure of curing, but construction schedules and economical reuse of forms often dictate removal of forms before the end of the required curing time. If this occurs, one of the moisture retention methods that follow can be adopted. They include a wet covering such as burlap, a sprayed membrane compound, or impervious paper or plastic covering. Plastic film should not be used where surface appearance is critical, because it will result in mottled concrete coloring, although that will not affect the quality of the surface.

Most precast, and particularly prestressed concrete, other than decorative panels, is cured by means of accelerated curing procedures. Accelerated curing is achieved with the use of saturated steam or dry heat, which requires the concrete member to be sealed to prevent loss of mixing water.

Accelerating admixtures

Some specifications call for accelerating admixtures, either calcium chloride or nonchloride accelerating admixtures, as a means of increasing the rate of hydration and speeding the strength gain of concrete, particularly in cold weather. This reduces the time that protection is needed. Accelerating admixtures also tend to accelerate the setting and hardening of concrete, in some cases expediting the finishing operations. However, when offsetting cold temperatures, accelerating admixtures should only be used to maintain a setting time and bleed period consistent with that of the mixture during normal placement temperatures. When the air is moist, it may take a long time for the bleed water to evaporate, thus delaying finishing of the accelerated mixture. Once the bleed water is gone, floating and troweling should be quick; the results of the accelerator are soon evident.

Calcium chloride and other accelerators should be used only when absolutely necessary, and only with the approval of the engineer/architect. Side effects, such as corrosion of metals and increased susceptibility to sulfate attack and alkali-aggregate reaction may occur when calcium chloride is used. In warm weather, accelerators often result in rapid set of the concrete, making finishing difficult or impossible. Many people wrongly believe that calcium chloride will act as an antifreeze in concrete. With the maximum permissible dosage – 2 percent by mass of the cement – it cannot significantly lower the freezing temperature of concrete. Calcium chloride should never be used in prestressed concrete. Codes prohibit its use there because it may contribute to the corrosion of prestressing tendons.

Keeping moisture in the concrete

Concrete should be kept from drying long enough to reach adequate strength. Normally when concrete is placed, it contains more water than is needed for proper hydration. However, evaporation may reduce the retained water below what is needed for development of desired properties. Excessive evaporation can be prevented either by adding more water or by retaining the water already in the concrete. There are several acceptable ways to complete each of these jobs, several methods are discussed here.

Membrane-forming curing compounds

Curing compounds are a practical and widely used method for sealing moisture in concrete. Liquid membrane-forming curing compounds consist of waxes, natural and synthetic resins, and solvents. When properly applied to concrete, they form a moisture-retaining film. Sometimes white or gray coloring is added to curing compounds to reflect sunlight and show that the compound is being applied uniformly (Fig. 6.4). Disappearing dyes can also be used to show uniformity of application. These compounds should meet ASTM C309, "Standard Specification for Liquid Membrane-Forming Compounds for Curing Concrete," requirements for moisture retention. ASTM C309 includes a minimum moisture retaining capability of the curing compound when tested in accordance with ASTM C156, "Standard Test Method for Water Loss [from a Mortar Specimen] Through Liquid Membrane-Forming Curing Compounds for Concrete." This test method measures the moisture retention capability of the curing compound on a wood float finish. If the finish is modified, say to a trowel finish, the test is not measuring the moisture retention of the product, but the moisture retention of the finish. Products that do not retain moisture should not be considered curing compounds.

Curing compounds are applied at a uniform rate by brush, spray, or roller. Except on deeply textured surfaces, a single application is sufficient if it is thick enough to assure

Fig. 6.4—Membrane-forming curing compound should be applied to flatwork immediately after final finishing of the concrete. Here, a worker uses pressure spraying equipment to apply a smooth, evenly distributed coat (photo courtesy of PCA)

forming a continuous film. When feasible, two applications at right angles to each other are suggested to assure complete coverage. Generally, a coverage rate of 1 gallon for every 150 to 200 ft^2 (1 L for every 3.5 to 5 m^2) is needed. For paving work, twice these amounts may be required, depending on the material used and surface texture of the pavement. When curing compound is applied to formed surfaces, the concrete should be kept moist after form removal. A uniformly damp appearance with no free water is desired when the compound is applied. For best results on flatwork such as floors and slabs, the membrane compound should be applied after finishing and as soon as the free-water sheen has disappeared from the surface. Do not walk on the slab before the membrane compound has dried. Footprints may have a lasting detrimental effect.

Precautions: For flatwork, cover the dried curing compound with building paper or other material to protect it from damage until curing is complete. Do not apply curing compound to surfaces scheduled to receive more concrete, paint, tile, glue down carpet, or overlays that require a positive bond, unless you can be sure by test or experience that it will not interfere with bond or appearance of the covering material. In these cases, use the moist curing procedures described in this chapter.

Waterproof paper or plastic film

Waterproof paper meeting ASTM C171, "Standard Specification for Sheet Materials for Curing Concrete," requirements is made of two layers of kraft paper bonded together with bituminous material and reinforced with fibers. It is a practical way to retain moisture when curing concrete slabs. Delivered in rolls, waterproof paper lies flat when rolled onto a concrete slab. Edges should be lapped and sealed with waterproof tape (Fig. 6.5), and the paper should be weighted down to maintain contact with the concrete and to prevent wind damage. Where appearance is important, be sure that the paper is nonstaining. Also be aware that waterproof paper, like plastic film, may produce

Fig. 6.5—Waterproof paper is placed on a slab immediately after finishing is complete. Sheets should be lapped and edges weighted down to avoid wind damage (photo courtesy of PCA)

Fig. 6.6—Plastic sheeting used to protect fresh concrete (photo courtesy of PCA)

a mottled appearance of the concrete surface if the paper is wrinkled or not lying flat on the surface of the concrete.

Plastic film for curing, also addressed by ASTM C171, should be at least 4 mils (0.004 in. (0.10 mm)) thick and can be clear, black, or white (Fig. 6.6). White film is good for hot weather because it reflects the sun's heat, while black is favored in cold weather because it absorbs solar heat. Plastic sheets are not as heavy as waterproof paper and sizes tend to be larger. They wrinkle instead of lying flat, and they frequently leave the concrete surface with a patchy or mottled discoloration. Therefore, they are not recommended where appearance is important. The plastic film should be weighted down just as the waterproof paper is. Plastic film placed over wet burlap is a good curing material. However, the burlap should be thoroughly wetted to prevent it from absorbing water from the concrete. The more absorptive the material, the heavier it gets when saturated and the easier it is to keep from wrinkling and resulting in the mottled discoloration. There are also curing blankets available that combine thin plastic with an absorbent layer, like disposable diapers. These curing sheets tend to overcome the mottled discoloration affect common with other curing methods, and also leave the surface without a potential bond-breaking coating.

Water spray or soaker hose

Continuously spraying or sprinkling water from a fog nozzle is a good curing method if the concrete is kept continuously wet (Fig. 6.7) and the excess water does not result in any other complications to the project, such as oversaturation of the subgrade. For vertical or sloping surfaces, soaker hoses can be used. Water pressure and flow should be low enough to avoid washing

Fig. 6.7—Soaker hoses can be used to keep fresh concrete wet for curing (photo courtesy of PCA)

away the new surface. Be sure that the water is nonstaining where appearance is important in the completed structure. Never allow the concrete surface to dry between sprinklings; this can cause surface crazing of the surface and cracking of slabs.

Wet burlap or mats
Wet burlap, cotton mats, rugs, or other absorbent coverings will hold water on either horizontal or vertical concrete surfaces. Burlap should be washed and free of anything that could stain or harm concrete, and it should be thoroughly saturated before applying it to the concrete (Fig. 6.8). The burlap or mat should be either kept damp with repeated applications of water (Fig. 6.9) to replace the loss from evaporation, or it should be covered with plastic. Cotton mats and rugs hold water longer and present less risk of drying out, but because they are heavier than burlap, the concrete surface should be stronger before they can be safely applied.

Other methods
Damp earth, sand, sawdust, straw and hay are also recommended for curing, but they have practical disadvantages. Earth and sand are messy, and hard to handle and to clean up. Straw and hay can dry out quickly, blow away, and can be fire hazards. When these materials are used for curing, they should be kept continuously wet. Ponding is sometimes used for slabs-on-ground, flat roofs, pavements, or wherever a pond of water can be created by a ridge; dike of earth, or other material at the edge of the slab; or where there is a stream of water, as through a culvert. Curing water should be no more

Fig. 6.8—Wet burlap can be used to help retain moisture to cure fresh concrete (photo courtesy of PCA)

Fig. 6.9—When burlap or other absorbent materials are used for concrete curing, they should be kept continuously wet during the entire curing period (photo courtesy of PCA)

than about 20°F (11°C) cooler than the concrete to prevent thermal stresses that could cause cracking.

Formwork: Wood forms, kept wet, and metal forms provide some protection against loss of moisture. Exposed top surfaces of the concrete should be kept wet to assure that the water runs down the inside of the previously loosened forms, if necessary, to keep the concrete wet. Otherwise, forms should be removed as soon as possible, so that the prescribed curing can begin with the least delay after placing. If a curing compound is to be used for surfaces on which the forms have been left for 24 hours or more, the surface should be soaked before curing compound is applied.

Cold weather precautions

Maintaining a concrete temperature above 40°F (4°C) is important to promote hydration, the most important precaution for concreting in cold weather is to protect the concrete from freezing at early ages, at least until its compressive strength is 500 psi (3.5MPa) or greater. At 50°F (10°C), most well-proportioned mixtures reach this strength after 24 to 48 hours. Concrete receiving initial protection from freezing will eventually mature to its potential strength despite later exposure to cold weather. ACI 306R recommends air-entrainment for any concrete that is exposed to freezing-and-thawing cycles during construction, even though it may not be so exposed in service. However, air-entrained concrete should not be given a hard troweled finish as there is risk of surface delamination.

It is common to require that a specified strength be attained within a few days or weeks, because of the need for efficiency of construction cycles, or to put the structure into service. Because concrete at low temperatures sets slowly and has delayed strength development, various ways to warm and maintain the warmth have been developed. It is important to recognize that insulating blankets typically only retain heat and, without an external source of heat, do not increase the concrete temperature. Blankets placed on cold concrete will not increase the temperature unless they are warmed by sun exposure or a built-in heat source.

ACI 306R recommends protection that maintains a concrete temperature above 50°F (10°C) for timely strength development in winter weather. Except within heated enclosures, little or no added moisture is required for curing during cold weather. At times the air inside heated enclosures becomes dry enough to cause plastic shrinkage cracking.

Job planning for cold weather should include one or more of the following recognized protective measures:

- Heating the water and concrete materials
- Heating the area in which the concrete is placed
- Employment of extra cement or high-early-strength cement
- Addition of an accelerating admixture to the mixture

Use vented heaters that do not give off carbon dioxide gas, because this can cause dusting of the hardened concrete.

Form surfaces should be at least a few degrees above freezing to avoid endangering surfaces and edges of the newly placed concrete. All snow, ice, and frost should be removed so that they do not occupy space intended to be filled with concrete. Concrete should not be placed on frozen subgrade material, and shoring should not be supported from a frozen subgrade. If the subgrade has to be thawed before placement, it may also have to be recompacted.

Fig. 6.10—Protective winter enclosure made of plastic sheeting. Clear grades of this material admit substantial amounts of natural light (photo courtesy of PCA)

Protection against freezing

Protection against freezing during the first day or two is very important. Massive structures such as bridge piers, dams, and mat foundations can be protected by insulating the forms and covering the top surface with insulation. These heavy sections tend to generate enough heat from cement hydration, provided insulation is in place to retain it. In very cold weather the form surfaces should be preheated.

Thinner sections are harder to protect. Several types of insulation materials are suitable for or specially produced for formwork:

- Sheets of polystyrene or polyurethane foam that may be cut to shape and fitted between the framing members of vertical formwork

Fig. 6.11—An enclosure with a heater provides cold weather protection. Combustion gases should be vented outside the enclosure (photo courtesy of PCA)

- Urethane foam sprayed onto the outside of forms, making a continuous layer of insulation
- Pliable vinyl foam blankets with an extruded vinyl backing; these are sometimes made with embedded electrical heating elements
- Mineral wool or cellulose fibers, usually protected within a heavy waterproof casing to form large mats, bats, or rolls

ACI 306R explains how much insulation to use, and how long it should stay in place.

The best assurance against concrete freezing, according to ACI 306R, is to work within heated, weather-tight enclosures. They are kept warm by low pressure steam or electric or vented fuel-burning heaters. However, the hot, dry air should not blow directly on the new concrete, otherwise rapid drying and shrinkage will occur. Within the enclosure, the impervious materials or curing compounds described above should be used on exposed surfaces. Water application is not advised in cold weather. Combustion chambers of fuel-burning space heaters should be sealed and vented to the outside atmosphere, or they will cause carbonation of the concrete surface, a condition that leads to dusting later. Although generally more expensive than other protection, heated enclosures also protect workers and improve their productivity.

At the end of the curing period, artificial heating should be discontinued and housings removed in such a manner that the concrete surface temperature drops uniformly, not more than 50°F (28°C) in 24 hours for sections thinner than 12 in. (300 mm), nor more than 20°F (11°C) in 24 hours for massive sections (72 in. (1.8 m) or more), as recommended by Table 5.1 of ACI 306R. These permissible temperature changes should

Fig. 6.12—Insulation blanket designed primarily for use on concrete slabs has been wrapped around this column form and secured in place (photo courtesy of David W. Johnston)

be reduced for concrete slabs-on-ground, whose thermal shrinkage is restrained by subgrade friction. When the temperature falls too rapidly, there is a great risk of cracking.

Hot weather precautions

Hot weather may lead to problems in mixing, placing, and concrete curing that result in poorer long-term strength and serviceability. Wind, sun, and dry air can aggravate the effects of high concrete temperature and high air temperature. Most problems relate to an increased rate of cement hydration at higher temperatures, and to faster evaporation of moisture from the freshly mixed concrete.

Improper curing during extreme hot weather can cause nonuniform surface appearance and low concrete strength that result from excessive and rapid evaporation. Increased plastic shrinkage cracking (see "Plastic shrinkage cracking" section) is also possible. Concrete placed at temperatures above 70°F (21°C) will require more water to maintain slump, may undergo premature stiffening, and may have relatively lower strength at later ages. Figure 6.13 shows how higher concrete curing temperatures give better 1-day strength results, but have the opposite effect on 28-day strengths.

ACI 305R recommends procedures to avoid trouble with concrete in hot weather. Some of those include:
- Keeping the surroundings cool
- Cooling the concrete and keeping it cool
- Working rapidly to avoid delays

In addition, mixture proportions can be adjusted to compensate for some problems. Retarding admixtures (ASTM C494/C494M) help offset the higher water demand that accompanies elevated temperatures. High-range, water-reducing admixtures that produce flowing concrete (ASTM C494/C494M and

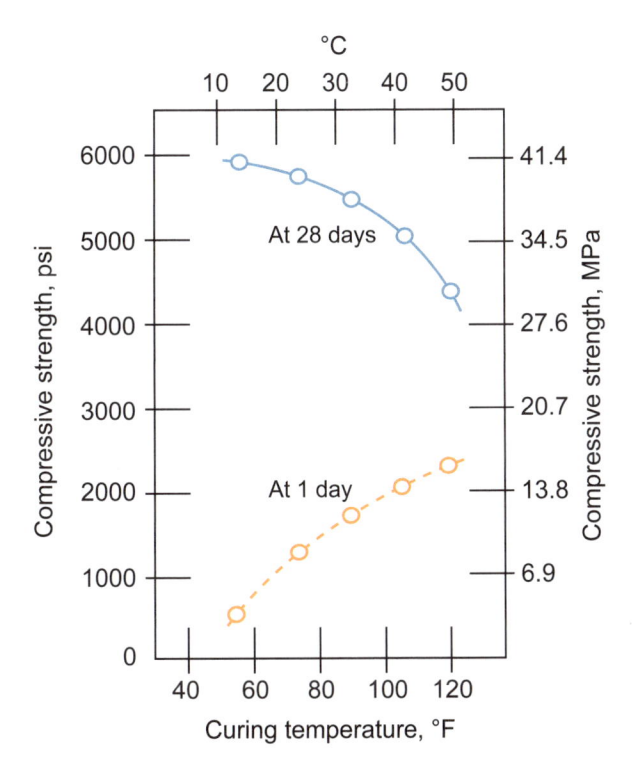

Fig. 6.13—Increasing the curing temperature gives better 1-day strength, but causes lower 28-day strength for the same concrete mixture (ACI 305R)

ASTM C1017/C1017M), popularly called superplasticizers as discussed earlier in this book, offer significant benefits during hot weather.

Keeping cool

Before concrete is placed, periodic spraying of forms, reinforcing steel, and subgrade with water will keep them cool and prevent absorption of water from the concrete. After concrete is in the forms, periodic spraying of the outside will help it cool and cure properly. Tops of walls, columns, and other vertical elements should be wet-cured and covered.

As the concrete temperature goes up, it requires more water to obtain a given slump, with other factors being equal. Concrete temperature can be effectively lowered by cooling the ingredients, especially water. Using ice as part of the mixing water or adding shaved ice directly to the mixer will help lower temperatures. Liquid nitrogen can be injected into a holding tank to chill mixing water, or it can be injected into the mixing water stream, turning the water to slush as it is discharged into the batch. Liquid nitrogen can also be injected to cool mixed concrete in trucks or central mixers (Fig. 6.14). Aggregate temperature is frequently lowered by sprinkling and shading from the sun or occasionally by refrigeration.

Avoiding delays

Ready mixed concrete trucks should be carefully scheduled so they can discharge promptly on arrival at the site. This avoids slump loss that may come with prolonged mixing and from trucks kept standing in the sun. Painting the trucks white or other light colors also helps keep the concrete cool. In extremely hot areas, some builders schedule night placement of the concrete. This not only takes advantage of lower temperature, but also saves transit time because traffic is usually lighter.

In extremely hot and dry weather, protection during placing and finishing may be needed. Windbreaks, plastic sheeting, fog sprays, evaporation retarders, or other protection may be needed to prevent drying during finishing. Starting curing immediately

Fig. 6.14—Liquid nitrogen can be used as a cooling method for hot weather concreting (photo courtesy of PCA)

CHAPTER 6—CURING AND PROTECTION

after finishing is extremely important during hot weather. Either apply a moisture retaining material or use one of the wet curing methods. If a membrane-forming compound is used, use white to reduce heat absorption from the sun. Ponding of water on the surface, where practical, is favored by some authorities.

Plastic shrinkage cracking

Cracks that appear on unformed surfaces soon after concrete placement, while the concrete is still plastic, are usually caused by excessive evaporation due to extreme heat and dry wind. They are commonly called plastic shrinkage cracks. Plastic shrinkage cracks may also occur in winter, particularly in the dry air of heated enclosures as the evaporation rate is dependent on the air temperature, air relative humidity, concrete temperature, and wind speed. These cracks are usually randomly oriented and unconnected, wide at the surface but of only moderate depth. Fog spraying (Fig. 6.15), wind screens, sun shades, plastic sheeting covers, or other procedures to inhibit moisture loss between finishing operations will materially reduce plastic shrinkage cracking in flatwork placed under unfavorable dry conditions. An evaporation retardant, monomolecular film, can be sprayed

Fig. 6.15—Fogging a bridge placement (photo courtesy of PCA)

on the surface of the fresh concrete after screeding, and as needed during floating and troweling operations.

Sometimes, the plastic shrinkage cracking can be prevented or remedied by timely working of the surface, accompanied by a somewhat later than usual hand floating, followed by a slightly early troweling. Cracks that have formed before troweling can often be beaten together with the float. If merely troweled over, the cracks are likely to show later.

CHAPTER 7—FIELD TESTING AND CONTROL OF CONCRETE QUALITY

Concrete control tests are generally performed by ACI Certified Field Testing Technicians but concrete craftsmen should be familiar with these tests and understand what the results mean.

Some tests are performed to determine if the concrete meets the job specifications. These tests are called "acceptance" tests because if the concrete fails to meet the specifications it can be rejected. Because such tests determine whether concrete should be accepted or rejected, they are required to be performed precisely as specified in the test standards. For example, the standard method for performing a slump test requires the use of a special steel tamping rod. If a piece of rebar or a wooden rod is used instead, the results of the test are not valid and may result in rejecting the concrete.

Most of the control tests for concrete have been standardized by ASTM International. Such tests are identified by a number, year and a title; an example ASTM C39/C39M-15a, "Standard Test Method for Compressive Strength of Cylindrical Concrete Specimens." The last two digits, "15" in this example, indicate that the standard was issued or revised in 2015. The final letter, "a" indicates more than one revision in the given year. Test methods are periodically revised, and each year ASTM publishes a complete volume of standards; these are also available online. Those for concrete and aggregates appear in Volume 04.02 (ASTM 2015). Several ASTM specifications are also included in the ACI *Field Reference Manual* (SP-15).

Sampling fresh concrete (ASTM C172/C172M)

Concrete used for control tests should be representative of the entire batch. For ready mixed concrete and concrete from stationary mixers, two or more samples are taken from the middle of the batch, with no more than 15 minutes between taking the first and the last sample. The sample size should be at least 1 ft^3 (0.03 m^3) if strength test specimens are to be made. Individual samples are combined and remixed with a shovel. Tests for slump and air content should be started within 5 minutes of obtaining the last portion of the sample. Molding specimens for strength tests should begin within 15 minutes of mixing the composite sample. If concrete is pumped, additional checking at the point of deposit may be needed because pumping can change the concrete slump or air content.

Fig. 7.1—Sampling ready mixed concrete delivered to the jobsite is performed in accordance with ASTM C172/C172M, "Standard Practice for Sampling Freshly Mixed Concrete" (photo courtesy of Frances Griffith)

CHAPTER 7—FIELD TESTING AND CONTROL OF CONCRETE QUALITY

Fig. 7.2—Sampling concrete from a pump

Slump test (ASTM C143/C143M)

A slump test is used to measure the consistency of concrete. In concrete mixes, changes in slump most often reflect changes in the amount of water or aggregate properties in the mixture, or changes in temperature, hydration, and setting. Thus, the slump test is an indication of the placeability of the mixture. Maintaining a consistent setting time and bleed period that results in consistent finishability is as important as achieving a consistent finish. The order of finishing normally follows that of placement so variation can impact the overall finish.

The essential equipment consists of a standard slump cone, a standard tamping rod, a scoop, and a ruler. A slump cone is made of sheet metal and is 12 in. (300 mm) high with a 4 in. (100 mm) diameter opening at the top and 8 in. (200 mm) diameter opening at the bottom. The tamping rod is a straight steel rod of 5/8 in. (16 mm) diameter and 24 in. (600 mm) long with the tamping end rounded to a hemispherical tip. The scoop can be of any convenient size.

Fig. 7.3—Slump test equipment: slump cone, tamping rod, base plate, ruler, scoop, and a timer

To perform a slump test, dampen the inside of the cone and place it on a rigid, flat, damp, and nonabsorbent horizontal surface. Fill the cone with concrete in three layers of approximately equal volume (Fig 7.4), rodding each layer 25 times with the tamping rod. For the bottom layer, the rod should penetrate to the bottom of the layer. The tamping rod should penetrate through each new layer and roughly 1 in. (25 mm) into the previous layer. After the three layers have been rodded, strike off and smooth the concrete at the top, and clean away excess concrete around the cone base. Then slowly lift the slump cone vertically without shaking or twisting, taking approximately 5 seconds to raise the cone 12 in. (300 mm). Immediately place the cone upside down next to the slumped concrete, and measure the vertical distance to the nearest 1/4 in. (5 mm), between the top of the cone and the center of the displaced top surface. If the concrete decidedly falls away to one side, disregard the

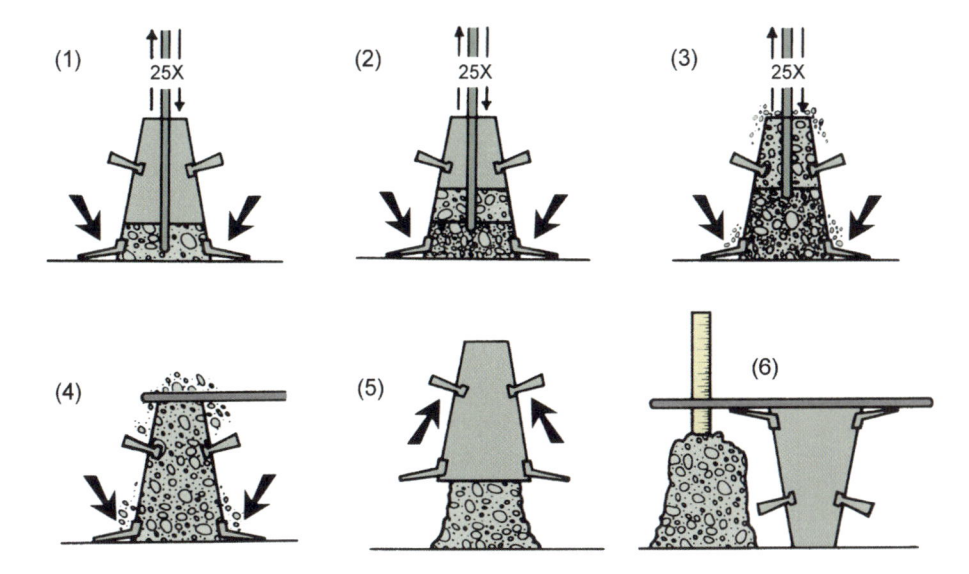

CHAPTER 7—FIELD TESTING AND CONTROL OF CONCRETE QUALITY

Fig. 7.4—Measuring the slump of fresh concrete. The cone is filled with concrete in three layers of equal volume (Steps 1, 2, and 3). Each layer is rodded 25 times with a steel tamping rod. After the top surface is smoothed (Step 4), the slump cone is slowly lifted vertically (Step 5) and turned upside down. The slump is measured (6) as the distance that the center of the top surface has settled

test and make a new test. If the concrete in the new test also decidedly falls to one side, then the concrete is probably not cohesive enough to have a meaningful slump.

Although a change in slump usually means a change in the amount of water in the mixture, it could also indicate other changes including those to air content, aggregate gradation, sand content, temperature, hydration rate, setting time, or misdosed admixtures.

The result of a single slump test should not be the basis for rejection of concrete because the test is subject to considerable variation, particularly if the testing technicians are not adequately trained. For example the slump may be too much if the base is subjected to jarring, or too little if the base is rough or dry. If the required slump is stated as a single number, say 5 in. (125 mm), a tolerance of ± 1 in. (25 mm) is normally considered acceptable; that is, the slump could be from 4 to 6 in. (100 to 150 mm). Frequently, specifications give the maximum permitted value, such as, "the slump shall not exceed ___ in." In this case, a lower slump, up to 2-1/2 in. (65 mm) less, may be acceptable, but slumps greater than the value given are not permitted. Refer to ACI 117 for additional details and other tolerances for concrete.

Air content tests

Air-entrained concrete contains many extremely small air bubbles. These bubbles are so small that there can be millions of them in a cubic inch (25 x 25 x 25 mm) of air-entrained concrete that contains 4 to 6 percent air. These air bubbles act as frictionless ball bearings in fresh concrete, thereby improving its workability. More importantly, these air bubbles improve the concrete's resistance to damage from freezing-and-thawing cycles. If hardened concrete is wet and then freezes, moisture in the concrete expands as it turns to ice. As the moisture expands, it is forced to move through the pores in cement paste. If the moisture has to move very far, for example, more than 0.008 in. (0.2 mm), pressures high enough to crack the concrete can occur. The entrained air bubbles act as empty chambers into which the expanding water can flow, thus relieving the destructive pressures.

Air content of air-entrained concrete should be checked on a regular basis because too little air will not provide resistance to freezing-and-thawing and too much air will result in low strength.

The pressure meter is useful for most types of concrete but it can give misleading results for some lightweight concretes and for other concretes which contain porous aggregates.

Both the pressure method and the volumetric method are somewhat complicated and time-consuming. Very often, a precise measurement of air content is not required and making an estimate is useful. In such cases, either a *unit weight test* or an *air indicator* test can be made. The air indicator test, however, has not been standardized and should not be used to determine compliance with specification limits, according to ACI 212.3R, "Report on Chemical Admixtures for Concrete."

Changes in air content for a specific mixture may result from variation in mixture components, such as the amount and type of air-entraining admixture, sand, or cement. Changes in air content can also result during placement, especially when a concrete pump is used.

Air content by the pressure method (ASTM C231/C231M)

A pressure meter (Fig. 7.5) works on the principle that a change in pressure of air results in a change in volume. The bowl of the air meter is filled with concrete and consolidated. If the concrete slump is less than 1 in. (25 mm), consolidation is performed by vibration and the bowl is filled in two equal layers. If the concrete slump is greater than 3 in. (75 mm), consolidation is performed by rodding and the

Fig. 7.5—Measuring the air content with a Type-B pressure meter (photo courtesy of Frances Griffith)

bowl is filled in three equal layers. If the slump is between 1 in. (25 mm) and 3 in. (75 mm), either method is acceptable. The concrete is struck off and smoothed, and the cover assembly is clamped onto the bowl. The test procedure is different for different types of meters. With one type of meter, water is added to fill the cover assembly up to a certain level. A cap then seals the assembly and air is pumped into the air space above the water until a certain pressure is reached. The water level goes down according to the air content of the concrete.

In another type of meter, air is pumped into a chamber to an initial pressure. A valve between the chamber and bowl is then opened and the drop in pressure indicates the air content of the concrete.

Air content by the volumetric method (ASTM C173/C173M)
This test, commonly referred to as the roll-a-meter test, works for all concretes, but it is used mostly for lightweight concrete. The air meter (Fig. 7.6) consists of a bowl and a top section. The bowl is filled with concrete in two layers of approximately equal volumes and each layer is rodded 25 times as for a slump test. The top layer is then struck off and

Fig. 7.6—Determining air content using a volumetric air meter (photo courtesy of Frances Griffith)

smoothed and the top section is clamped on. The top section is then carefully filled with water and isopropyl alcohol to the zero mark and the top cap is screwed on. The air meter is then rolled and agitated until all of the air has been removed from the concrete. The drop in the liquid level reveals the air content. If the liquid level does not stabilize within 6 minutes, or there is more foam than that equivalent to 2 percent air, the sample is required to be discarded and the test repeated with greater amount of isopropyl alcohol.

Air content estimated with an air indicator

An air indicator, Fig. 7.7, is a pocket size device. The rubber stopper contains a brass cup. To check the air content, fill the cup with a representative sample of mortar with no aggregate particles larger than 1/10 in. (2.5 mm) from the concrete, and insert the stopper and cup into the glass tube. Fill the tube up to the zero mark with alcohol and then with your thumb over the open end of the tube shake the tube to remove the air from the mortar (paste). Read the drop in the alcohol level and from a chart, the air content is estimated. This is not a standard test and should not be used to determine compliance with specification limits to accept or reject a load of concrete. This test is simple to run and generally gives a good approximation of the air content. To properly estimate the air content using the air indicator, you will need to know the mortar content of the concrete.

(a) Fill the open-top cup with mortar from the concrete. Be careful to avoid any pieces of gravel or rock.

(b) Insert the stopper, which contains the cup, into the glass tube.

(c) Fill the glass tube with alcohol to the zero mark.

(d) Shake the tube (thumb over the open end) until the mortar is mixed with the alcohol.

(e) The drop in alcohol level is a measure of the air content.

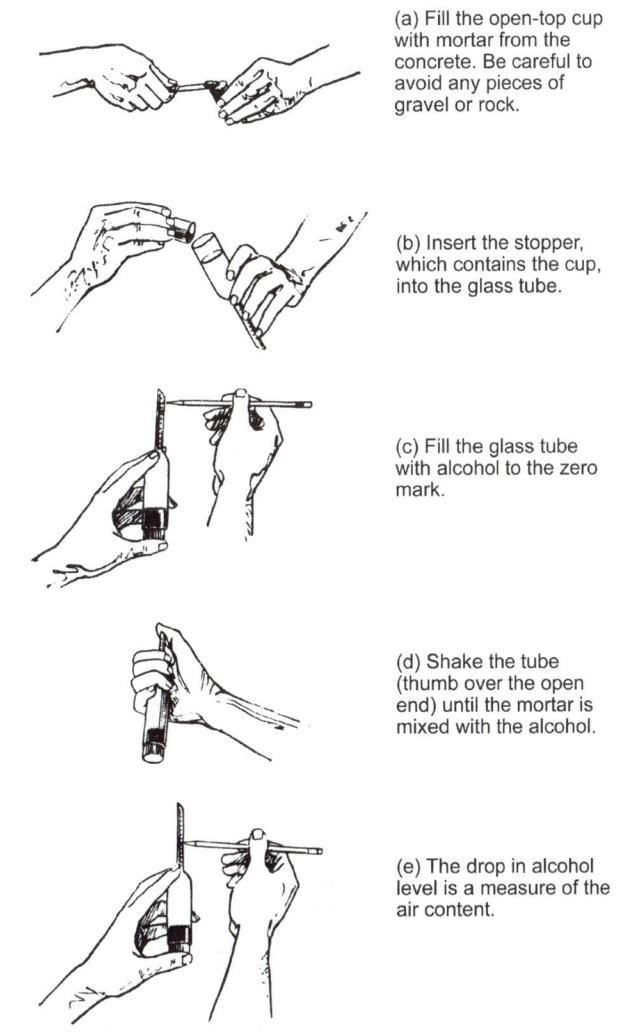

Fig. 7.7—Air content of fresh concrete estimated with an air indicator test

Density (unit weight) and yield (ASTM C138/C138M)

Yield is the volume of freshly mixed concrete produced from a mixture of known quantities of the component materials. Yield computations can be used to determine actual cement content, or to check the batch-count volume against the observed volume in place. If the total quantity of mixing water is obtained, the *w/cm* can be verified. ASTM C138/C138M uses the density (unit weight) method for determining yield.

This test is simple but requires an accurate scale and a sturdy cylindrical measure (Fig.7.8). To perform the test, fill the measure (cylindrical container) in either three or two equal layers, depending on the consolidation method. Consolidation by rodding requires three layers, while consolidation by vibration requires two layers. For concrete with slump greater than 3 in. (75 mm), consolidate concrete by rodding. For concrete with slump less than 1 in. (25 mm), consolidate concrete by vibration. For concrete with slump from 1 in. (25 mm) to 3 in. (75 mm), consolidate by either method. Consolidation by rodding requires rodding each

Fig. 7.8—To determine the density (unit weight) of concrete, record the mass of a level-full measure of concrete, subtract the mass of the empty measure, and divide by the volume of the measure (photo courtesy of Frances Griffith)

layer 25 times with a standard tamping rod (the same type used for a slump test). After consolidation, strike off the top surface, smooth it, and wipe the sides of the container clean. Then weigh the full container to the nearest 1/10 lb (45 g), or to within 0.3 percent of the measured mass, whichever is greater. Subtracting the mass of the empty measure gives the net mass of the concrete. Divide the net mass of the concrete by the volume of the container to get the density (unit weight) of the concrete.

Example:
Suppose that a cylindrical measure having a volume of 0.25 ft^3 (0.00708 m^3) and mass of 8.3 lb (3.8 kg) is filled with concrete as described previously. Suppose that the mass of the filled bucket is 44.7 lb (20.3 kg). The density (unit weight) of the concrete can be calculated as:

Example (in.-lb units):
Mass, container and concrete = 44.7 lb
Mass, container = 8.3 lb
Mass, concrete = 44.7 lb − 8.3 lb = 36.4 lb
Volume of measure = 0.25 ft^3
Density (unit weight) of concrete = 36.4 lb ÷ 0.25 ft^3
= 145.6 lb/ft^3

Example (SI units):
Mass, container and concrete = 20.3 kg
Mass, container = 3.8 kg
Mass, concrete = 20.3 kg − 3.8 kg = 16.5 kg
Volume of measure = 0.00708 m^3
Density (unit weight) of concrete = 16.5 kg ÷ 0.00708 m^3
= 2330 kg/m^3

Changes in density (unit weight) usually indicate changes in air content but may reflect changes in materials or mix proportions.

When the total mass of materials in a batch is known, the yield of the batch can be calculated by dividing the total mass by the density (unit weight) of the concrete. For example, if the total mass of cement, aggregates, water, and admixture in a batch is 31,450 lb (14,265 kg) and if the concrete density (unit weight) is 145.6 lb/ft^3 (2330 kg/m^3):

Yield = 31,450 lb ÷ 145.6 lb/ft^3 = 216 ft^3 (in.-lb units)

Yield = 14,265 kg ÷ 2330 kg/m^3 = 6.12 m^3 (SI units)

For in.-lb units only:
Divide 216.0 ft^3 by (27 ft^3/yd^3) = 8.00 yd^3
The batch yield is 8 yd^3 (6.12 m^3).

ASTM C138/138M also provides equations for calculating cement content and air content in the tested mixture.

Temperature (ASTM C1064/C1064M)

Temperature is a very important factor in concrete work. A standard method of measuring the temperature of concrete is provided by ASTM C1064/C1064M. To get accurate results, use a thermometer or other temperature measuring device verified to be accurate to plus or minus 1°F (0.5°C) throughout a range of 30°F to 120°F (0°C to 50°C). Insert the temperature measuring device into a representative sample of concrete to a depth of at least 3 in. (75 mm). Record the temperature after 2 minutes, but no more than 5 minutes after device insertion. An average of two or more readings is suggested.

Dial type metal thermometers (Fig. 7.9) with ranges between 30°F and 120°F (0°C to 50°C) are generally used for concrete. Infrared guns are also used to measure concrete temperature but be aware that many factors, including sunlight, angle of infrared beam, and concrete color, can have a significant impact on the temperature reading.

Fig. 7.9—A dial type metal thermometer is used to get the temperature of fresh concrete (photo courtesy of Frances Griffith)

Making test cylinders (ASTM C31/C31M)

Most specifications require compressive strength results of several cylinder tests to be averaged to give a true indication of the concrete strength. In addition, one or more *hold cylinders* may be called for as backups to be used in case the 28-day cylinders are damaged or do not come up to strength.

Procedures for testing and evaluating the test results are discussed in Chapter 8 of this book.

In the U.S., it is common to make concrete test cylinders that are either 4 in. (100 mm) diameter by 8 in. (200 mm) high (4 x 8 in. (100 x 200 mm)), or 6 in. (150 mm) in diameter and 12 in. (300 mm) high (6 x 12 in. (150 x 300 mm)) cylinders. If specifications follow ACI 318, each test requires two 6 x 12 in. (150 x 300 mm) cylinders, or three 4 x 8 in. (150 x 300 mm) cylinders. Thus, if tests are required at three different ages, at least six 6 x 12 in. (150 x 300 mm) cylinders or nine 4 x 8 in. (100 x 200 mm) cylinders are needed.

Fig. 7.10—Concrete test cylinder molds are filled with fresh concrete in three equal layers and, each layer is rodded 25 times with a tamping rod. Each cylinder should be labeled and covered with a plate or otherwise protected from moisture loss (photo courtesy of Frances Griffith)

There are several kinds of cylinder molds including heavy steel, sheet metal (tin can), plastic, and waxed cardboard molds. The tin can, cardboard, and some plastic molds can be used only once. The heavy steel mold and some sheet metal and plastic molds are reusable, so they should be cleaned and oiled after each use. Cylinder molds are required to comply with ASTM C470/C470M, "Standard Specification for Molds for Forming Concrete Test Cylinders Vertically."

After obtaining a representative sample of concrete (ASTM C172/C172M) as explained in this chapter, the procedure is to:

1. Place the cylinder molds, cleaned and oiled if required, on a solid level base such as a concrete slab.

2. Select the proper method for filling the molds based on the slump.

   a.) *For slumps less than 1 in. (25 mm):* Fill the mold in two equal layers. Use an internal vibrator by inserting it into the concrete at one, two, or three locations for 4 x 8 in. (100 x 200 mm), 6 x 12 in. (150 x 300 mm), or 9 x 18 in. (225 x 450 mm) cylinders, respectively. Leave the vibrator in the concrete long enough each time to allow entrapped air to escape, then raise the vibrator slowly, and tap the sides of the mold lightly.

b.) *For slumps equal to or greater than 1 in. (25 mm):* Consolidate the concrete by rodding or vibrating. If rodding, place concrete into the molds in two, three, or four layers for 4 x 8 in. (100 x 200 mm), 6 x 12 in. (150 x 300 mm), or 9 x 18 in. (225 x 450 mm) cylinders, respectively. Use 3/8 in. (10 mm) diameter tamping rod for 4 x 8 in. (100 x 200 mm) cylinders, or 5/8 in. (16 mm) diameter tamping rod for 6 x 12 in. (150 x 300 mm) or 9 x 18 in. (225 x 450 mm) cylinders. Rod each layer 25 times (Fig. 7.10) in case of 4 x 8 in. (100 x 200 mm) or 6 x 12 in. (150 x 300 mm) cylinder molds. Rod each layer 50 times in case of 9 x 18 in. (225 x 450 mm) cylinders molds.

3. Strike off and smooth the surface; the top surface of cylinders should be protected from moisture loss.

4. After initial curing, usually about 24 hours, and within 30 minutes after removing from molds, mark each cylinder so that it can be matched with the concrete in a particular part of the project, also record the time and date. Cure specimens with water maintained on their surfaces at all times at a temperature of 73.5°F ± 3.5°F (23.0°C ± 2.0°C) using water storage tanks or moist rooms.

Fig. 7.11—One method of protecting the top of concrete cylinders from moisture loss is to use a plastic cap (photo courtesy of PCA)

Curing and protecting test cylinders
Cylinders are made and tested for two reasons:
1. To determine if the concrete meets the specified compressive strength (design) requirements.
2. To determine if concrete, in place, has the strength needed to remove the forms or to put the concrete into service.

Cylinders for design strength check
Cylinders made for design strength check should be stored in a moist environment where the temperature is 60 to 80°F (16 to 27°C) for up to 48 hours. ASTM C31/C31M, "Standard Practice for Making and Curing Concrete Test Specimens in the Field," suggests several ways to maintain this satisfactory moisture and temperature. Then remove the cylinders from the mold and keep them moist at 70 to 77°F (21 to 25°C) until the time of test. This test is usually done in the lab. Tighter temperature requirements are prescribed for concrete cylinders representing concrete mixtures with specified strength of 6000 psi (40 MPa) or greater. If the cylinders are to be sent to a laboratory for standard curing before 48 hours, be sure that they remain in the molds and

Fig. 7.12—Concrete specimens (beams and cylinders) are stored in a moist environment meeting requirements of ASTM C511, "Standard Specification for Mixing Rooms, Moist Cabinets, Moist Rooms, and Water Storage Tanks Used in the Testing of Hydraulic Cements and Concretes," for final curing

are kept moist until they reach the lab. They should not be transported until at least 8 hours after final set. Transportation time should not exceed 4 hours. During transportation, specimens should be cushioned to prevent jarring-related damage and covered with materials that prevent moisture loss. Additional protection includes adding insulation materials during cold weather. When they reach the lab, they are demolded and placed in standard curing until the test time. Care in shipment and storage is essential to the achievement of accurate test results. Recognize that these strength tests measure the concrete's ability to meet the specified strength requirement and acceptance of the concrete mixture as delivered. The strength of the cylinders may not represent the strength of the concrete in-place, especially if the temperature and moisture-retention of the concrete in-place has not been kept equal to that of the cylinders. This can become critical in hot or cold weather where the cylinders are maintained at standard conditions but the in-place concrete is not.

Cylinders made for construction site control

Cylinders made for construction site control are stored differently than those made for design strength check. They should be kept at the jobsite temperature and given the same curing as the concrete they represent.

Specimens made to determine when a structure can be put into service should be removed from the molds at the time of removal of the formwork. They are tested in the moisture condition resulting from the jobsite storage. Careful handling and transportation for testing are very important for these cylinders too.

CHAPTER 8—EVALUATING CONCRETE STRENGTH

Core and Cylinder Strength Tests of Hardened Concrete

The standard method for determining concrete strength during construction consists of making and testing concrete compressive and flexural strength specimens. Laboratory-cured specimens are required to show that the concrete as delivered has an acceptable level of strength. The most common specimens are cylinders, made as described in Chapter 7 of this book and ASTM C31/C31M. They should be tested in compression in accordance with ASTM C39/C39M. Flexural strength tests of beam-shaped specimens are sometimes required, particularly in paving work.

Building officials or specifiers may also require tests of field-cured specimens to check on adequacy of curing and protection of concrete in the structure. Field-cured specimens may also be used, particularly in high-rise construction, to show that the structure is strong enough to permit the removal of formwork or shoring.

The moisture content of the specimen has considerable effect on the resultant strength. A saturated specimen will show lower compressive strength and higher flexural strength than companion specimens tested dry. This makes it important to follow the ASTM testing methods precisely if valid, accurate strength readings are to be obtained.

ACI 318 requirements

Section 26.12.2.1 of ACI 318 requires compression tests of laboratory-cured test cylinders for all structural concrete, except when the quantity of a given class of concrete is less than 50 yd^3 (38 m^3), provided the "evidence of satisfactory strength is submitted to and approved by the building official." Samples for making the test cylinders are required to be taken at least once a day, and not less than once for every 150 yd^3 (110 m^3) placed, and at least once for each 5000 ft^2 (460 m^2) of surface area for slabs and walls. Remember that the Code defines a strength test as the average strength of two 6 x 12 in. (150 x 300 mm) cylinders or three 4 x 8 in. (100 x 200 mm) cylinders made from the same sample and tested at the same designated age, usually 28 days.

Concrete strength is considered acceptable when the following two conditions are met:
1. The average of all sets of three consecutive strength tests equals or exceeds the specified compressive strength, f'_c.

2. No individual strength test, average of two 6 x 12 in. (150 x 300 mm) cylinders or three 4 x 8 in. (100 x 200 mm) cylinders, is more than 500 psi (3.5 MPa) below f'_c if f'_c is 5000 psi (35 MPa) or less, or more than $0.10f'_c$ if f'_c exceeds 5000 psi (35 MPa).

If either of these requirements is not met, changes in mixture proportions and placing practices shall be made to improve the strength for future batches. This can be applied only on jobs that go beyond the 28 days required to get the first strength test results.

Cylinder compressive strength tests (ASTM C39/C39M)

Compressive strength of concrete is generally measured using cylinder specimens prepared and tested in accordance with ASTM C39/C39M (Fig. 8.1). Cylindrical samples should have a diameter at least three times the maximum size of the coarse aggregate in the concrete and a length as close to twice the cylinder diameter as possible.

The ends of samples for compression testing should be ground or capped in accordance with the requirements of

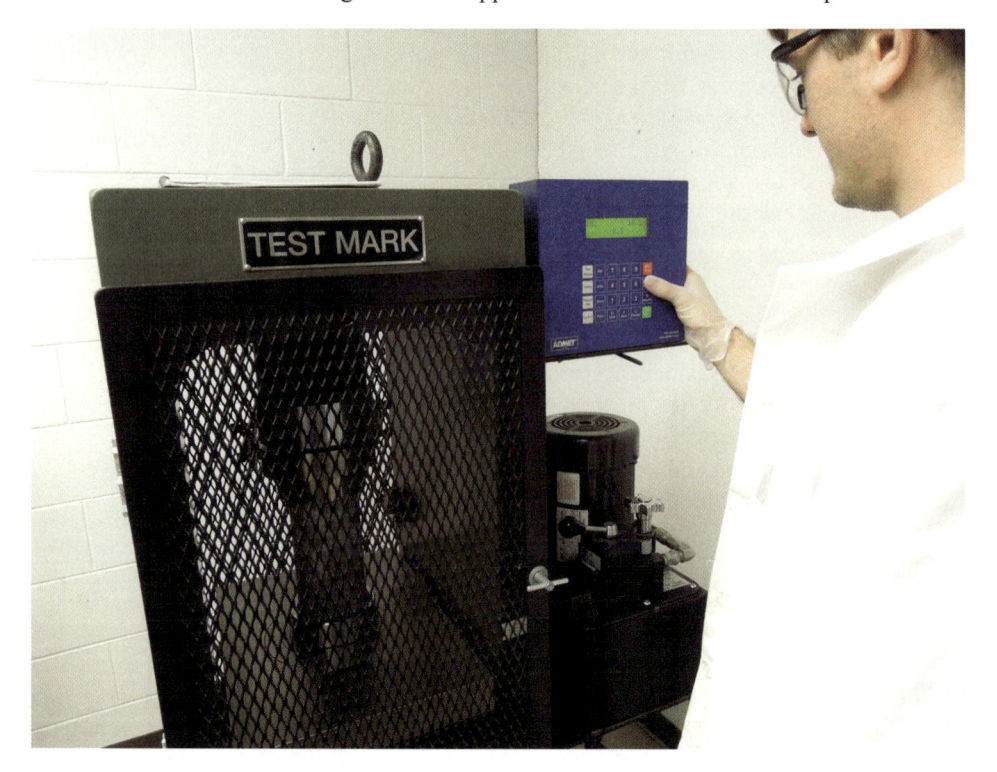

Fig. 8.1—A 6 x 12 in. (150 x 300 mm) concrete cylinder being tested in compression. Load is required to be applied at a uniform rate until the cylinder fails (photo courtesy of Frances Griffith)

ASTM C617/C617M, "Standard Practice for Capping Cylindrical Concrete Specimens." The end of the cylinder is required to be a smooth, plane surface to assure even loading and accurate test results. Various commercially available materials can be used to cap compressive strength test specimens. Proprietary or laboratory-prepared sulfur mortars can be used if the caps are allowed to harden at least two hours before the specimens are tested (Fig. 8.2). Caps should be made as thin as is practical.

Most job specifications require that concrete reach at least 3000 psi (21 MPa), 4000 psi (28 MPa), or some other minimum strength at 28 days. This specified compressive strength is commonly referred to as f'_c, and a 28-day strength test is always used unless specified otherwise. This means that the average compressive strength of cylinders made, cured, and transported in accordance with ASTM C31/C31M and tested in a standard manner (ASTM C39/C39M) needs to meet the specified strength.

Twenty-eight days after the cylinders are made they are placed in a calibrated power operated testing machine that

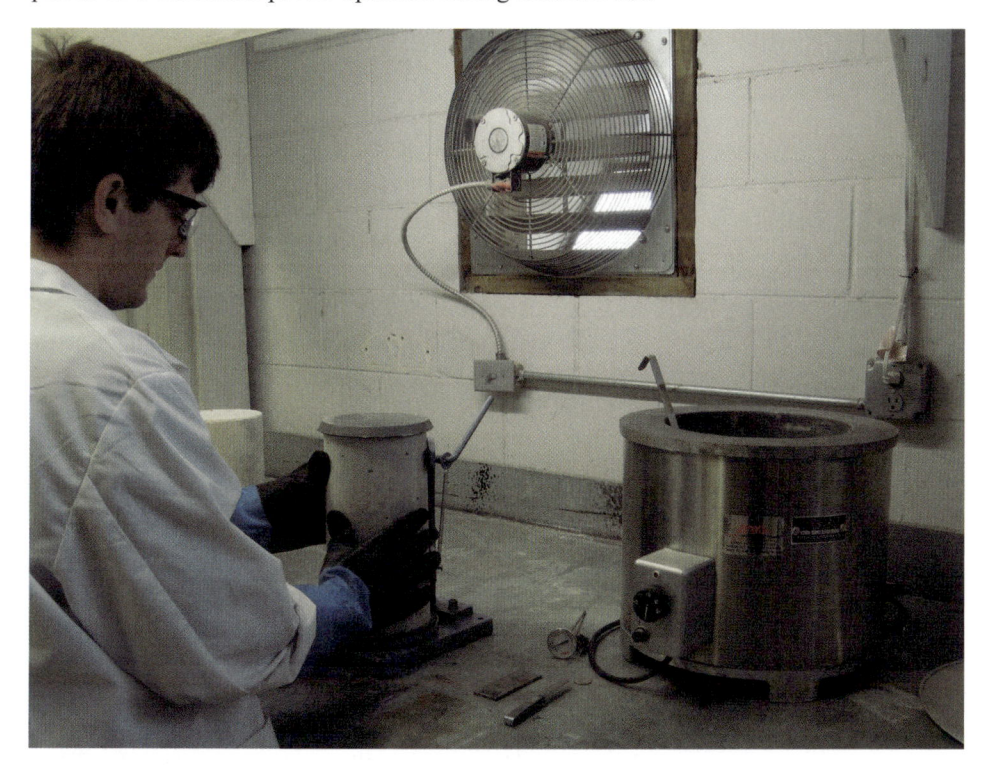

Fig. 8.2—Capping test cylinders with sulfur compound before testing. Properly made caps permit even loading across the cross section of the cylinder (photo courtesy of Frances Griffith)

applies load at a uniform rate to the flat ends. The load is increased until the cylinder fails under load. Divide the maximum load by the area of the flat surface, to get concrete strength. For example if the cylinder is 6 x 12 in. (150 x 300 mm), and the maximum load is 90,000 lb (400 kN), the strength is:

$$\text{strength} = \frac{\text{maximum load}}{\text{load area}}$$

The loaded area for a 6 in. (150 mm) diameter circle is 28.3 in.^{2} (18300 mm^2), so:

$$\text{strength} = \frac{90{,}000 \text{ lb}}{28.3 \text{ in}^2} \text{ (in.-lb units)}$$

$$= 3180 \text{ lb/in.}^2 \text{ (psi)}$$

$$\text{strength} = \frac{400 \text{ kN}}{18300 \text{ mm}^2} \text{ (SI units)}$$

$$= 21.9 \text{ MN/m}^2 \text{ (MPa)}$$

Core tests (ASTM C42/C42M)

When the probability of low strength concrete in the structure is suspected by unsatisfactory cylinder test results or other means, the building official or the design engineer may require tests of cores drilled from the structure. Concrete cores do not always result in a sample with height-to-diameter ratio of 2. For samples with lengths of 1 to 2 times the diameter, correction factors found in ASTM C42/C42M, "Standard Test Method for Obtaining and Testing Drilled Cores and Sawed Beams of Concrete," shall be used. Cores with a height of less than 95 percent of the diameter before or after capping should not be tested.

Cores should not be taken until the concrete can be sampled without disturbing the bond or adhesion between the paste and the coarse aggregate. For horizontal surfaces, cores should be taken vertically and not near formed joints or edges. For vertical or sloped faces, cores should be taken perpendicular to the central portion of the concrete element. Coring through reinforcing steel should be avoided. A cover meter (electromagnetic device) can be used to locate steel. Cores taken from structures that are normally wet or moist in service should be moist conditioned and tested moist, as described in ASTM C42/C42M. Those from structures normally dry in service should be conditioned in an atmosphere approximating their service conditions and tested dry.

If the structure is dry in service, prior to testing, the cores should be dried for 7 days at a temperature of 60 to 80°F (16 to 27 °C) and a relative humidity of less than 60 percent. The cores should be submerged in water for at least 40 hours before testing if the structure is to be used in a more than superficially wet state.

Cores should be inspected for cracks before testing. The type of break is evaluated after testing to determine whether a test should be included in the average. If the average strength of three cores is at least 85 percent of f'_c and if no single core is less than 75 percent of f'_c, the concrete in the area represented by the core is considered structurally adequate. If the results of properly made core tests are so low as to leave structural integrity in doubt, strength evaluation as outlined in Chapter 27 of ACI 318 may be necessary.

Nondestructive and in-place testing methods

Nondestructive and in-place test methods are valuable aids in the overall quality assurance for large projects. However, they cannot be used to evaluate strength until comprehensive laboratory studies have correlated traditionally tested strength values for the field materials and mixture proportions being used with field readings obtained using the nondestructive method. Without these correlations, the various nondestructive tests can only serve to evaluate the relative strengths of hardened concrete. The most widely used in-place tests are the rebound, penetration, pullout, and dynamic or vibration tests. They should all be performed by an accredited testing laboratory.

When approved by the building official, some of these tests are appropriate to demonstrate the necessary structural strength required for form removal.

Both the rebound hammer and the probe damage the concrete surface to some extent. The rebound hammer leaves a small indentation on the surface; the probe leaves a small hole and may cause minor cracking and small craters similar to popouts. The pullout test leaves a cone-shaped void that should be repaired.

Rebound hammer test (ASTM C805)

Figure 8.3 shows a workman using an impact rebound hammer, frequently referred to as a Schmidt hammer. It consists of a spring-controlled hammer that slides on a plunger (Fig. 8.4). When the plunger is pressed against a concrete surface, the hammer retracts against the force of the spring. When the hammer is completely retracted, the spring is automatically released (Fig. 8.4c). The hammer impacts the shoulder area of the plunger, and the spring-controlled mass rebounds (Fig. 8.4d). The rebound distance is measured on a scale attached to the instrument. This distance is called the "rebound number."

The results of a Schmidt rebound hammer test (ASTM C805/C805M, "Standard Test Method for Rebound Number of Hardened Concrete") are affected by surface smoothness, size, shape, and rigidity of the specimen; age and moisture condition of the concrete; type of coarse aggregate; and carbonation of the concrete surface. When these limitations are recognized and the hammer is calibrated for the particular materials used in the concrete, this instrument can then be useful for determining the relative compressive strength and uniformity of concrete in the structure. Even though the rebound numbers are not a precise indication of the concrete strength, higher numbers mean greater strength when comparing concretes made of the same materials and cast at about the same time.

Use of the impact rebound hammer seems simple but it is easy to get misleading readings.

Fig. 8.3—An impact rebound hammer gives an indication of the concrete hardness. The device is useful for finding areas of poor quality (photo courtesy of PCA)

CHAPTER 8—EVALUATING CONCRETE STRENGTH

Penetration resistance method (ASTM C803/C803M)

Like the rebound hammer, this is basically a hardness tester that provides a quick means of determining the relative strength of the concrete. The equipment specified by ASTM C803/C803M, "Standard Test Method for Penetration Resistance of Hardened Concrete," consists of a powder-actuated driver that drives a hardened alloy probe or pin into the hardened concrete (Fig. 8.5). The penetration resistance is determined by measuring the exposed length of the probes driven into the concrete or by measuring the depth of the hole created by the pin penetration into the concrete.

The results of the penetration test are influenced by surface smoothness of the concrete and the type and hardness of aggregate used. For this reason, a calibration table or curve is needed for the particular concrete to be tested, usually from cores or cast specimens, if the goal is to estimate compressive strength from the test.

Because the area tested by the probe or pin is small, there is more variability in results than would be expected from cylinder tests. However, it is effective for measuring relative rate of strength development of concrete at early ages, especially for determining stripping time for formwork.

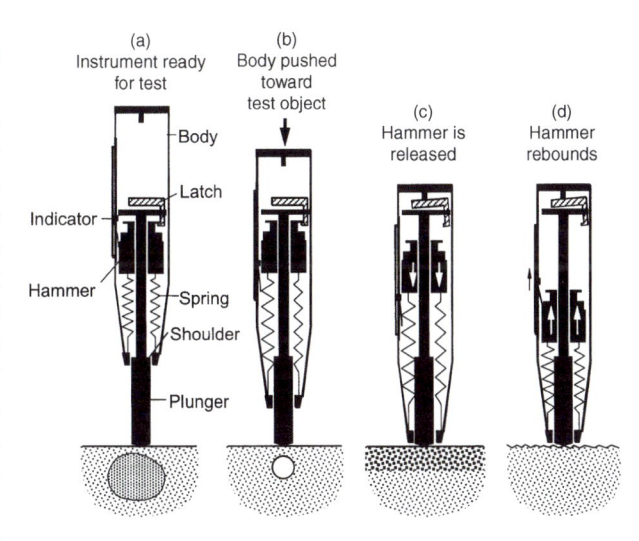

Fig. 8.4—Schematic illustration shows how the impact rebound hammer works. Note that the hammer itself is inside the body of the instrument. Surface or near-surface condition of the concrete, such as a hard aggregate particle (a), large air void (b), surface carbonation (c), or rough surface (d) can influence the reading (from ACI 228.1R, "In-Place Methods to Estimate Concrete Strength")

Fig. 8.5—For the penetration resistance test, a powder actuated gun drives a hardened probe or pin into the concrete. The operator measures the depth of hole or exposed length of the pin (photo courtesy of PCA)

Pullout tests (ASTM C900)

The pullout test (ASTM C900, "Standard Test Method for Pullout Strength of Hardened Concrete") consists of pulling out of the hardened concrete a specially shaped insert whose enlarged end has been cast into the fresh concrete (Fig. 8.6 and 8.7). A dynamometer measures the force required to pull it out. Because of the shape of the insert, a cone of concrete is pulled out with it leaving a hole to be patched.

The test provides a measure of the shear strength of the concrete that can be converted to equivalent compressive strength, using correlation and calibration data specific to the concrete mixture and test insert being used. The pullout test is used to monitor early-age strength development of concrete in order to determine appropriate form stripping times.

Unlike other in-place tests, the pullout test has to be planned in advance. ASTM C900 sets requirements for the number, location, and spacing of pullouts when they are to be the basis for form removal.

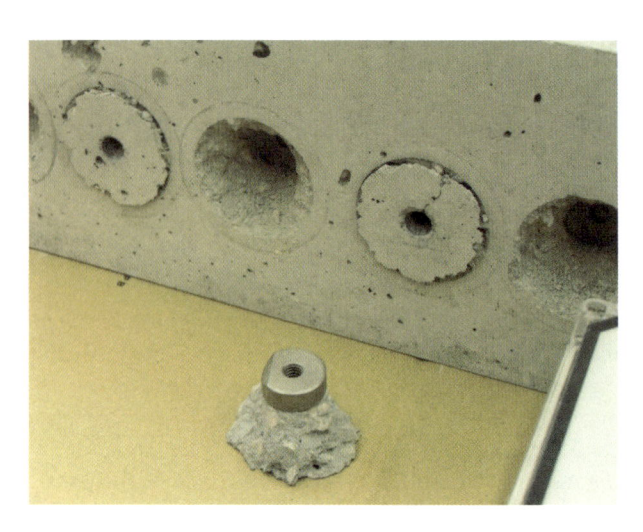

Fig. 8.6—Pullout test instrument pulls out a metal insert that was set in place before the concrete was cast. A dynamometer measures the required force (photo courtesy of David W. Johnston)

Pulse velocity test (ASTM C597)

The ultrasonic pulse velocity method (ASTM C597) consists of measuring electronically the travel time of an ultrasonic wave through concrete. High velocities generally indicate sound concrete and low velocities indicate unsound concrete. The method is now used only for comparing relative strengths and detecting nonuniformities in the concrete.

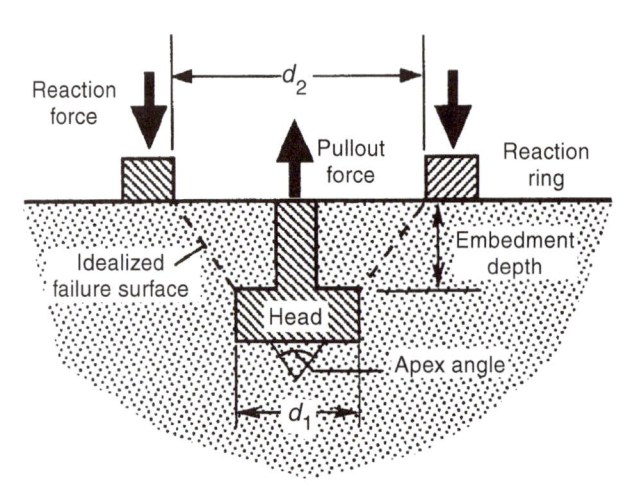

Fig. 8.7—Schematic of pullout test. A roughly cone-shaped piece of concrete is removed during the test (from ACI 228.1R)

Concrete maturity method (ASTM C1074)

Concrete temperature varies over time due to several factors including the ambient temperature, the heat generated by the chemical reaction between the water and cementitious materials, and the insulating properties of the formwork. The maturity method (ASTM C1074, "Standard Practice for Estimating Concrete Strength by the Maturity Method") uses recordings of the concrete temperature over time to estimate concrete compressive strength.

A concrete laboratory will establish a correlation between compressive strength and the area, A, between the concrete temperature, T_t, and the reference temperature, T_0. This area is shown in Fig. 8.8. The correlation testing is performed on the specific concrete mixture that will be used in the field and is not applicable to other concrete mixtures.

In the field, concrete temperature versus time is recorded for the concrete placed in the structure. The shape of the field concrete temperature versus time curve will be different than the laboratory curve. The difference is accounted for by using the area, A. This area is used with the correlation established by the laboratory to estimate the compressive strength of the field concrete at a particular age.

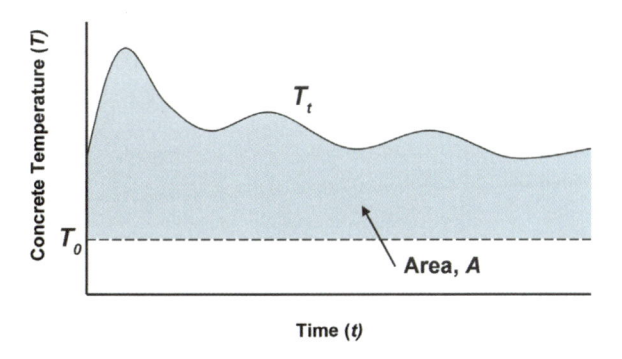

Fig. 8.8—Example of concrete temperature versus time curve for use with the maturity method

The temperature history of the concrete in the field is typically measured by embedding several thermocouples in the concrete. Temperature readings from the thermocouples are tracked over time using a data logging system, some of which have been specifically designed for use with the maturity method. Products (Fig. 8.9) have also been developed that can be embedded in the concrete and calculate maturity (area under the temperature curve) directly and the data transmitted to a readout or recording device.

Load testing concrete structures

If there is doubt about the safety of a structure or part of a structure, the engineer or the building official may order load tests. This might happen if some of the materials supplied were considered deficient in quality, if the construction methods were suspected of being substandard, or if the structure failed to meet some of the building code requirements.

Fig. 8.9—Temperature and maturity data loggers positioned in formwork prior to concrete placement (photo courtesy of Anthony P. Gallodoro)

Load tests are not made until after the structure or portion of it in question is 56 days old, unless owner of the structure, the contractor, the licensed design professional and all other involved parties agree to an earlier test age. Details of the test procedure and interpretation are generally placed under the direction of an engineer experienced in structural investigation, field tests, and measurements.

According to ACI 318 the test load should be applied in at least four approximately equal increments and left in place for 24 hours, unless signs of distress are observed. Tanks of water or stacks of heavy construction materials are sometimes used to provide the load. The load is removed immediately after all response measurements are recorded.

If the structure is to be considered acceptable, it should not show any visible evidence of failure, such as cracking, spalling, and large deflections. If the structure shows no visible evidence of failure, then recovery of deflection after removal of the test load is used to determine whether or not the strength of the structure is satisfactory.

Refer to Chapter 27 of ACI 318 for details of strength evaluation by load test, including test load arrangement, load factors, test load application, response measurements, and acceptance criteria.

APPENDIX A—REFERENCES

1. Kosmatka, S. H., and Wilson, M. L., 2011, *Design and Control of Concrete Mixtures*, fifteenth edition, Portland Cement Association, Skokie, IL, 444 pp.

2. ACI Committee 311, 2007, Manual of Concrete Inspection, ACI SP-2, American Concrete Institute, Farmington Hills, MI, 196 pp.

3. Portland Cement Association, 1971, "Concrete for Small Jobs," IS174.04T, Skokie, IL, 8 pp.

4. National Ready Mixed Concrete Association/American Society of Concrete Contractors, "Checklist for the Concrete Pre-Construction Conference," Joint NRMCA/ASCC publication, 18 pp.

5. Contractor's Guide to Quality Concrete Construction, third edition, 2005, joint publication of American Concrete Institute, Farmington Hills, MI, and American Society of Concrete Contractors, St. Louis, MO, 147 pp.

Referenced standards and committee reports

The following lists show the ACI and ASTM standards and reports that are cited in this book. These documents are revised from time to time, generally in minor detail. If you want to get a complete copy of the current version, contact the publishers at these addresses:

American Concrete Institute
38800 Country Club Drive
Farmington Hills, MI 48331-3439
USA
+1.248.848.3700

American Society for Testing and Materials
100 Barr Harbor Drive
P.O. Box C700
West Conshohocken, PA 19428-2959
USA
+1.877.909.2786

American Concrete Institute

ACI 117-10 Specification for Tolerances for Concrete Construction and Materials and Commentary (Reapproved 2015)

ACI 211.1-91 Standard Practice for Selecting Proportions for Normal, Heavyweight, and Mass Concrete (Reapproved 2009)

ACI 212.3R-10 Report on Chemical Admixtures for Concrete

ACI 228.1R-03 In-Place Methods to Estimate Concrete Strength

ACI 301-10 Specifications for Structural Concrete

ACI 304R-00 Guide for Measuring, Mixing, Transporting, and Placing Concrete (Reapproved 2009)

ACI 305R-10 Guide to Hot Weather Concreting

ACI 306R-10 Guide to Cold Weather Concreting

ACI 308R-01 Guide to Curing Concrete (Reapproved 2008)

ACI 309R-05 Guide for Consolidation of Concrete

ACI 318-14 Building Code Requirements for Structural Concrete and Commentary

ACI SP-15(10) Field Reference Manual: Specifications for Structural Concrete ACI 301-10 w/Selected ACI & ASTM References

ACI SP-4(14) *Formwork for Concrete*

CCS-1(10) Slabs-on-Ground

CCS-2(84) Cast-in-Place Walls

CCS-3(89) Supported Beams and Slabs

CCS-4(08) Shotcrete for the Craftsman

CCS-5(16) Placing and Finishing Decorative Concrete Flatwork

CT-13 ACI Concrete Terminology

E2-00(06) Reinforcement for Concrete

ASTM International

American Society for Testing and Materials, 2015, "Concrete and Aggregates," *Annual Book of ASTM Standards*, V. 04.02, ASTM International, West Conshohocken, PA, 1120 pp.

ASTM C31/C31M-15 Standard Practice for Making and Curing Concrete Test Specimens in the Field

ASTM C33/C33M-13 Standard Specification for Concrete Aggregates

ASTM C39/C39M-15a Standard Test Method for Compressive Strength of Cylindrical Concrete Specimens

ASTM C42/C42M-13 Standard Test Method for Obtaining and Testing Drilled Cores and Sawed Beams of Concrete

APPENDIX A—REFERENCES

ASTM C91/C91M-12 Standard Specification for Masonry Cement

ASTM C94/C94M-15a Standard Specification for Ready-Mixed Concrete

ASTM C136/C136M-14 Standard Test Method for Sieve Analysis of Fine and Coarse Aggregates

ASTM C138/C138M-14 Standard Test Method for Density (Unit Weight), Yield, and Air Content (Gravimetric) of Concrete

ASTM C141/C141M-14 Standard Specification for Hydraulic Hydrated Lime for Structural Purposes

ASTM C143/C143M-15 Standard Test Method for Slump of Hydraulic-Cement Concrete

ASTM C150/C150M-15 Standard Specification for Portland Cement

ASTM C156-11 Standard Test Method for Water Loss [from a Mortar Specimen] Through Liquid Membrane-Forming Curing Compounds for Concrete

ASTM C171-07 Standard Specification for Sheet Materials for Curing Concrete

ASTM C172/C172M-14a Standard Practice for Sampling Freshly Mixed Concrete

ASTM C173/C173M-14 Standard Test Method for Air Content of Freshly Mixed Concrete by the Volumetric Method

ASTM C191-13 Standard Test Method for Time of Setting of Hydraulic Cement by Vicat Needle

ASTM C231/231M-14 Standard Test Method for Air Content of Freshly-Mixed Concrete by the Pressure Method

ASTM C260/C260M-10a Standard Specification for Air-Entraining Admixtures for Concrete

ASTM C309-11 Standard Specification for Liquid Membrane-Forming Compounds for Curing Concrete

ASTM C470/C470M-15 Standard Specification for Molds for Forming Concrete Test Cylinders Vertically

ASTM C494/C494M-15 Standard Specification for Chemical Admixtures for Concrete

ASTM C511-13 Standard Specification for Mixing Rooms, Moist Cabinets, Moist Rooms, and Water Storage Tanks Used in the Testing of Hydraulic Cements and Concretes

ASTM C595/C595M-15e1 Standard Specification for Blended Hydraulic Cements

ASTM C597-09 Standard Test Method for Pulse Velocity Through Concrete

ASTM C617/C617M-15 Standard Practice for Capping Cylindrical Concrete Specimens

ASTM C618-15 Standard Specification for Coal Fly Ash and Raw or Calcined Natural Pozzolan for Use in Concrete

ASTM C685/C685M-14 Standard Specification for Concrete Made by Volumetric Batching and Continuous Mixing

ASTM C803/C803M-03(2010) Standard Test Method for Penetration Resistance of Hardened Concrete

ASTM C805/C805M-13a Standard Test Method for Rebound Number of Hardened Concrete

ASTM C845/C845M-12 Standard Specification for Expansive Hydraulic Cement

ASTM C900-15 Standard Test Method for Pullout Strength of Hardened Concrete

ASTM C989/C989M-14 Standard Specification for Slag Cement for Use in Concrete and Mortars

ASTM C1017/C1017M-13e1 Standard Specification for Chemical Admixtures for Use in Producing Flowing Concrete

ASTM C1064/1064M-12 Standard Test Method for Temperature of Freshly Mixed Hydraulic-Cement Concrete

ASTM C1074-11 Standard Practice for Estimating Concrete Strength by the Maturity Method

ASTM C1157/C1157M-11 Standard Performance Specification for Hydraulic Cement

ASTM C1602/C1602M-06 Standard Specification for Mixing Water Used in the Production of Hydraulic Cement Concrete

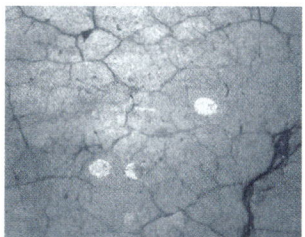

Custom Seminars

Personalized training to fit your organization's needs and goals

- Convenient
- Cost-effective
- Expert instructors
- State-of-the-art publications

www.ConcreteSeminars.com

aci UNIVERSITY

- Over 100 online courses
- Purchase through ACI Store
- Available 24/7
- Certificate programs

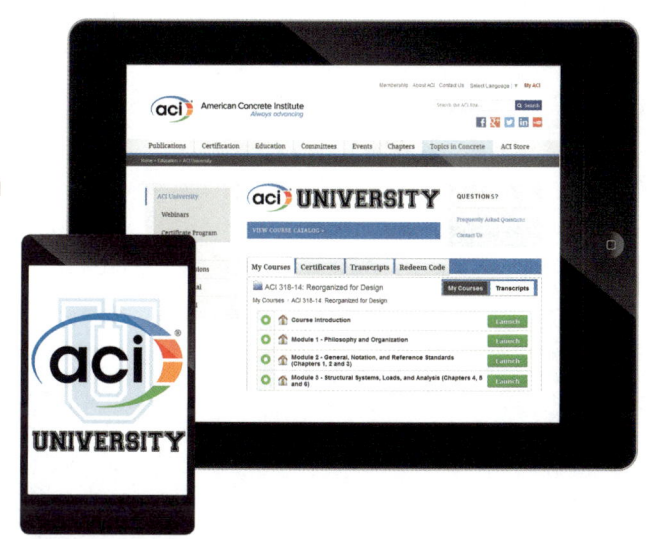

www.ACIUniversity.com

Get Certified

Since 1980, ACI has tested over 400,000 concrete technicians, inspectors, supervisors, and craftsmen in 20 different certification programs.

When you have a need for qualified concrete professionals—specify ACI Certification.

Visit www.ACICertification.org for:

Descriptions of ACI Certification Programs — Includes program requirements and reference/resource materials.

Schedule of Upcoming/Testing Sessions — Search by program and/or state.

Directory of Certified Individuals — Confirm an individual's certification and date of expiration.

Contractor's Guide to Quality Concrete Construction

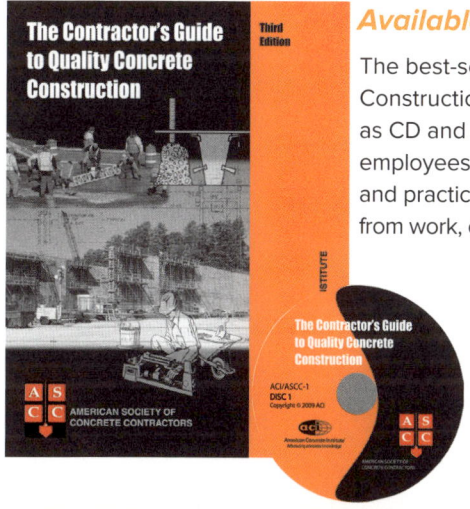

Available in text and audiobook versions.

The best-selling "Contractor's Guide to Quality Concrete Construction" is available in Spanish text format, as well as CD and MP3 audio formats. Educate yourself or your employees on quality concrete construction techniques and practices while waiting in a vehicle, traveling to and from work, or running between projects.

Formats:
Text
6-CD set
MP3
(Both audio formats include a 75-page printed book of photos, figures, tables, and checklists.)

www.concrete.org/store

Formwork for Concrete
Completely revised and updated; still the formwork reference of choice

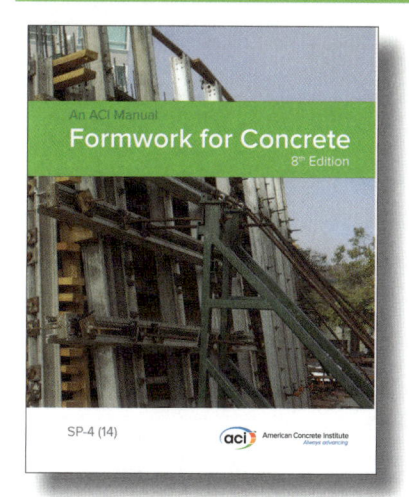

- **An ACI best-selling document**
- **Allowable strength design and load and resistance factor design examples**
- **Updated to current standards**
- **Chapter problems for classroom study**
- **500 modern color photographs**
- **150 color illustrations**
- **Includes ACI 347R-14**

www.concrete.org/store